Publisher's Message for
DYNAMIC FOREST

To shed light on today's cultural, social, economic, and political issues that are shaping our future as Canadians, Dundurn's **Point of View Books** offer readers the informed opinions of knowledgeable individuals.

Whatever the topic, the author of a **Point of View** book is someone we've invited to address a vital topic because their frontline experience, arising from personal immersion in the issue, gives readers an engaging perspective, even though a reader may not ultimately reach all the same conclusions as the author.

Our publishing house is committed to framing the hard choices facing Canadians in a way that will spur democratic debate in our country. For over forty years, Dundurn has been "defining Canada for Canadians." Now our **Point of View Books**, under the direction of general editor J. Patrick Boyer, take us a further step on this journey of national discovery.

Each author of a **Point of View** book has an important message and a definite point of view about an issue close to their heart. Some **Point of View Books** will resemble manifestos for action, others will shed light on a crucial subject from an alternative perspective, and a few will be concise statements of a timely case needing to be clearly made.

But whatever the topic or whoever the author, all these titles will be eye-openers for Canadians, engaging issues that matter to us as citizens.

*J. Kirk Howard*
President

T0125396

# A Note from the General Editor

A tree's leaf symbolizes our country on the national flag. Since Canada's centennial, Gordon Lightfoot's potent imagery about "the green dark forest" has moved us with primordial feelings about the dense woodlands that are both background and living presence in our lives. National income in the billions and jobs for many thousands come from exporting timber and wood products.

Forests come with issues, too. As well as enduring trade disputes with the Americans over softwood lumber, there have been determined battles between loggers and "tree huggers" about such things as felling the world's tallest Sitka spruce in B.C.'s Carmanah Valley, logging the West Coast's wilderness Stein River Valley, and harvesting in Ontario's provincial parks.

The saga of our country is entwined with forests. However, the reality of our national life — including economic, social, cultural, and environmental dimensions — being inextricably linked with forests does not mean we've got a clear fix on just what this relationship is, or should be.

Given that, it's clear greater knowledge and awareness are needed. *Dynamic Forest* offers a welcome guide to the forest and our relationship with it. Malcolm Squires's informative and engaging account reveals the intrinsic dynamism of forests, as well as the evolving outlooks of those living in harmony with them. Forester Squires scrutinizes the shortcomings of views ardently held, upholds the strength of Canadian democracy in shaping public policy, and demonstrates how the quest for "balance" in forestry policy and practice must account for the fact that, in nature, nothing is static or unchanging. He moves us beyond "logging" to forest management. He reminds us that without plants we can't continue to exist, but they can do very well without us. He takes us into his natural home — the boreal forest — and shows how lessons from Newfoundland and northwestern Ontario provide guidance for public policy and private action in the great boreal expanse from our Pacific to Atlantic coasts.

With "dirt under the nails" wisdom drawn from years of wide experience, Malcolm Squires makes a compelling case, applicable to far more than forestry, for avoiding destructive polarization and for seeing how knowledge promotes respect.

*J. Patrick Boyer*
General Editor
Point of View Books

# DYNAMIC FOREST

## Other Point of View Titles

*Irresponsible Government*
by Brent Rathgeber
Foreword by Andrew Coyne

*Time Bomb*
by Douglas L. Bland
Foreword by Bonnie Butlin

*Two Freedoms*
by Hugh Segal
Foreword by Tom Axworthy

*Off the Street*
By W.A. Bogart
Foreword by Sukanya Pillay

*Charlie Foxtrot*
by Kim Richard Nossal
Foreword by Ferry de Kerckhove

*Sir John's Echo*
by John Boyko
Foreword by Lawrence Martin

# DYNAMIC FOREST

## Man Versus Nature
## in the Boreal Forest

## MALCOLM F. SQUIRES

### Foreword by John Kennedy Naysmith

## DUNDURN
A J. PATRICK BOYER BOOK
TORONTO

Cover Image: Malcolm F. Squires
Printer: Webcom

**Library and Archives Canada Cataloguing in Publication**

Squires, Malcolm F., author
        Dynamic forest : man versus nature in the boreal forest / Malcolm
F. Squires; foreword by John Kennedy Naysmith.

(Point of view)
Includes bibliographical references.
Issued in print and electronic formats.
ISBN 978-1-4597-3932-1 (softcover).--ISBN 978-1-4597-3933-8 (PDF).--
ISBN 978-1-4597-3934-5 (EPUB)

1. Taigas. 2. Taiga conservation. 3. Taiga ecology. 4. Forest management. 5. Sustainable forestry. I. Title.
II. Series: Point of view (Dundurn Press)

SD387.T34S68 2017                333.75                C2017-903386-7
                                                       C2017-903387-5

1   2   3   4   5       21   20   19   18   17

  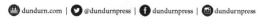 

We acknowledge the support of the **Canada Council for the Arts**, which last year invested $153 million to bring the arts to Canadians throughout the country, and the **Ontario Arts Council** for our publishing program. We also acknowledge the financial support of the **Government of Ontario**, through the **Ontario Book Publishing Tax Credit** and the **Ontario Media Development Corporation**, and the **Government of Canada**.

Nous remercions le **Conseil des arts du Canada** de son soutien. L'an dernier, le Conseil a investi 153 millions de dollars pour mettre de l'art dans la vie des Canadiennes et des Canadiens de tout le pays.

Care has been taken to trace the ownership of copyright material used in this book. The author and the publisher welcome any information enabling them to rectify any references or credits in subsequent editions.
                                                    — *J. Kirk Howard, President*

The publisher is not responsible for websites or their content unless they are owned by the publisher.

Printed and bound in Canada.

VISIT US AT

dundurn.com  |  @dundurnpress  |  dundurnpress  |  dundurnpress

Dundurn
3 Church Street, Suite 500
Toronto, Ontario, Canada
M5E 1M2

*To Muriel, who has shared my love for her*
*with my love for the forest*

*The wisdom in nature is distinguished
from that in man by the co-instantaneity
of the plan and the execution, the
thought and the product are one, or
are given at once ...*
— Samuel Taylor Coleridge

# CONTENTS

# FOREWORD

This book, *Dynamic Forest*, is the product of an experienced forester's perspective on a significant area of Canada's vast boreal forest, which in turn is typical of boreal forests indigenous to northern regions around the world.

The deep understanding that Mac Squires acquired of the boreal forest is the product of more than six decades of close association with it. An association that began as a boy growing up in a forest community in Newfoundland, where he spent much of his time roaming the forest and, as he says, "observing it going about its business." With that background, it was natural that he would choose to study forest science at university. A choice which in turn led to a very successful career as a professional forester, one characterized by his first-hand understanding of the boreal forest community.

The business Mac refers to stems largely from the connectedness of trees, water, air, and soil. A recent study of Canada's boreal forest suggests that the collective value of that connectedness, when quantified in terms of carbon storage, flood control, and water filtering, is in the order of $700 billion annually. By comparison, the average annual value of renewable and non-renewable products from Canada's boreal forest that flow through the market place is roughly 10 percent of that figure.

Going about its business, to be sure, and that study pertains only to Canada and does not include any of the boreal forests elsewhere in the world. Twenty-first century forests, considered on a global scale, have become invaluable.

A feature of Mac Squires' career in recent years has been his ability and willingness to communicate with the public on a broad spectrum of forestry topics, ranging from understanding how the forest works, to forest management strategies and initiatives directed at ensuring healthy forests in the future. An example of his success at reaching out to the larger community is the series of some sixty articles that he wrote over a three-year period for publication in a northwestern Ontario daily newspaper.

The considerable response that the published columns elicited ranged from appreciation for improving the readers' understanding of the natural forces inherent in the forest to thanks for helping readers gain an informed view regarding the effects of human intervention and subsequent efforts to maintain productive forest land. Not all readers supported the nature and extent of the latter. In Mac's view, that was fine. He felt that the purpose of the articles was to facilitate productive discussion based on correct information. That was happening.

One concern expressed by many who read Mac's columns was that the value of these newspaper articles was being limited as they were being read by a relatively small number of readers. It was frequently suggested that if they were published in book form, the potential for reaching a substantial number of readers could be attained. The book was seen as a step toward increasing society's awareness of the importance of the globe's forests. We are fortunate indeed that Dundurn Press has done just that.

The philosopher and poet George Santayana once said, "The earth makes music for those who listen." Now sit back and enjoy learning about some of the music Mac has heard during a lifetime of listening.

*John Naysmith R.P.F. (Ret'd) is a Lakehead University Fellow and recipient of a Charles Bullard Fellowship from Harvard. In his five-decade-long career, he served as a company logging superintendent in northern Ontario; chief of the Water, Forest and Land Division, Yukon and NWT; founding chair of Ontario's Forestry Futures Trust Committee; founding dean of Lakehead University's Faculty of Forestry; and as Canada's representative on the United Nations Advisory Committee on Forestry Education.*

# INTRODUCTION

The boreal forest is constantly changing, often dramatically. We like to picture the boreal forest as a stable, balanced system. It is anything but stable. Balanced? Yes, but only temporarily and in limited areas. It is resilient.

Over sixty years, progressing through bush worker, forestry student, practising forester, and retired independent thinker, I have seen forests that were "protected" within national parks become devastated by insects, moose, wind, and wildfire.[1] I have worked in and studied forests that over 110 years have been twice clear-cut harvested and are now mature again on private and public land. In all cases, the forest has returned; sometimes that forest is quite different from the original, and sometimes it is quite similar to the original, but never is it exactly the same as the original.

For decades I have monitored stands that were unaffected by dramatic depletion, such as that associated with clear-cut harvesting, only to discover that they had almost completely changed in tree and other plant species and wildlife habitat. I have become convinced that a naturally balanced boreal forest is a human concept that does not exist in nature. The boreal forest is always changing; the boreal forest is dynamic.

If we don't soon collectively recognize and accept that reality, and stop making what I feel are irrational demands that we tie up or "protect" ever-increasing areas of forests from change or human management, we may be conditioning the boreal forest for future disaster.

My love of that forest compelled me to write the articles on which *Dynamic Forest* is based. Readers of the articles convinced me I should write this book. In it I discuss the dangers I see if we follow the demands of those who want us to change the boreal forest from its natural, even-aged structure to an unnatural, uneven-aged structure. I end with a plea that we put aside our selfish wants and work together for the good of the forest, its inhabitants, ourselves, and our descendants.

# 1

## CANADA IS A FOREST NATION

Canada is a forest nation. Our forests benefit each and every one of us, regardless of whether we live in Whitecourt, Millertown, Gull Bay, or Toronto. Some of those benefits are obvious, but many are less obvious, especially if we live in larger cities far from the boreal forest. Those benefits are often thought of as separate and independent, but, like in the forest itself, every benefit is part of an interdependent whole.

The forest industry can't be easily separated from the transportation, energy, mining, and tourism industries, because to variable degrees they are all dependent on each other for their success. The success of businesses within those industries in turn helps ensure the economic and social viability of our communities and the quality of life we have come to expect.

There are many other benefits that, because of the wealth provided by business, we can participate in or use. First to come to mind are the numerous everyday products that are on our store shelves: lumber and the scores of paper and pulp products, such as tissue, magazine, packaging, and wrapping papers. Then there is wood fibre in some of our clothing and even forest-derived ingredients in some of our toothpastes and medicines. We keep adding new products with advancing knowledge and technology. As I write this, I am looking around my office and all, or a part, of practically everything I see came from the forest. The wood in the framing, walls, and floors of my house, the paper on the wall, my book case and the books in it, the reports and files in my filing cabinets, even some of the ingredients in the paints on the walls and in my art supplies are made to some degree from forest products.

All of those products are dependent on trees, but trees are only part of the forest. Think of the other products of the land: the wild meat, fish, skins, berries, and mushrooms. These are all items that we extract from the forest, and they easily command a dollar value. However, there are other, less tangible benefits, including carbon storage, oxygen production, water regulation, climate buffering, aesthetics, relaxation, and spiritual enhancement. The boreal forest is home to billions of migratory and non-migratory birds, mammals, insects, spiders, reptiles, amphibians, plants, and fungi. Time spent in the woods is beneficial to our mental and physical health. Yes, the boreal forest is a necessary part of our human habitat.

Of course, it is easier to quantify the economic importance of Canada's forest to Canadians. Our forests and forest industry supported directly and indirectly 288,669 jobs in 2014 and when the industry was at a low point it paid $8.657 billion in wages during 2012. Total revenue from goods manufactured by the industry was $53.159 billion in 2012 and total exports from the sector were valued at $30.774 billion in 2014.[1]

Because of the immense importance of forests to Canada and the world, it is imperative that we sustain them and ensure their health. In order to do that, it is necessary to use proper forestry practices, which in many cases, particularly in Canada's boreal forest, involves clear-cut harvesting. I acknowledge that probably the majority of Canadians believe that clear-cutting the forest is bad for the environment, but I intend to demonstrate in this book that if we are to live in, utilize, and sustain a healthy boreal forest, then clear-cut harvesting and even-aged stand management has to remain the predominant silvicultural system.

## SOME COMPARATIVE STATISTICS

Forests cover only 27 percent of the world's four billion hectare land area. Because of human population increase and forest area shrinkage, the available forest per person has dropped from 0.8 hectares in 1990 to only 0.6 hectares in 2015.[2] I feel that if that ratio continues to deteriorate we are headed for serious trouble. It is becoming clear to me that we must control our birthrate, sacrifice some of our wants, and co-operate at sustaining our forests.

Canada has a total of roughly 347 million hectares of forested area,[3] or approximately 9 percent of the world's total, and 9.5 hectares of forest for every one of its citizens. That provides us with a wealth of opportunity compared to the opportunity available to residents of most other nations. It also places a huge responsibility on each of us to ensure our forests are well managed.

Canada stands out in many ways among forested nations. For instance, only 6.2 percent of the Canadian forest is privately owned, and that is predominantly held by industry.

The provinces own 76.6 percent, the territories 12.9 percent, aboriginal 2.0 percent, and the federal government 1.6 percent.[4] The majority of forest harvesting in Canada is carried out by industry and their contractors through timber licences acquired from the provinces.

By comparison, in Finland, "individuals and families own some two-thirds of the Finnish forests."[5] "There are approximately 350,000 privately owned forests in Finland, making private, non-industrial forest owners central actors in the Finnish timber trade."[6] There, for practical reasons and economy of scale, land owners sell standing timber to firms that do the harvesting.

No two nations in the world have the same legislation, forest policies, land ownership patterns, climate, forest soil, plant and fauna species, and disturbance patterns. Even within Canada there are differences in all of those across the nation, sometimes over relatively short distances.

I have discovered in my reading that there is a trend among many nations, including in Canadian jurisdictions, toward returning forests to their more natural patterns. Most acknowledge that objective will never be fully achieved, because of our human need for resources.

Significantly, "more than 46 percent of the country's forests are certified. As of 2015, Canada had 166 million hectares of independently certified forest land. That represents 43 percent of all certified forests worldwide, the largest area of third-party certified forests in any country."[7] I am proud of that.

Fire is the greatest renewal agent in the Canadian boreal forest. The average area burned each year for the past twenty-five years has been approximately 2.5 million hectares, with this area being consumed by an average of approximately 7,500 fires.

Only 3 percent of the fires are responsible for burning 97 percent of the area that is burned,[8] doing so in large firestorms. In comparison, timber harvesting is responsible for only a fraction as much depletion of Canada's forests as that caused by fires. During 2015, the area harvested — 0.77 million hectares — was only 31 percent of that which is burned in the average year. Also in 2015, 0.44 million hectares, equivalent to 57 percent of the harvested area, was planted and 17.4 thousand hectares was seeded.[9] The remainder was left to regenerate naturally.

Among the Nordic and Baltic nations, "The highest proportion of planted forests, but also the smallest forest area, is found in Denmark, with an estimated 66 percent share of productive planted forests (FAO 2014b). Second and third are, however, the two largest countries, Sweden and Finland, with planted forest shares of 43 and 26 percent, respectively."[10]

## FOREST MANAGEMENT IS ABOUT BALANCING COMPETING INTERESTS

There are few easy decisions in forestry. Forest management is about balancing competing interests; an action that satisfies the needs of one person may, and usually does, damage the interests of others. Ontario's legislation and regulations that pertain to forest management require forest planners and practitioners to weigh competing interests and maintain balance while obeying the law. To be successful, forest managers have to be excellent dancers, or to use another analogy, they have to be excellent skaters and stickhandlers.

Competing interests include the province as it tries to satisfy our economic and social needs; First Nations' treaty rights;

forestry, tourism, mining, energy, and transportation industries; and cottagers, hunters, gatherers, those who find spiritual connection with nature, and others. Most of those interests are in direct conflict at some time over forest use, but sometimes they may co-operate when they perceive common ground.

Probably the conflict that the public is most aware of is that between First Nations and the forest industry. First Nations look to rights contained in their treaties with government, which they feel are being ignored by government and industry. Forestry companies look to government to guarantee a timber supply and the freedom to harvest using what they feel to be good forest management practices. For example, tensions have been high near the community of Grassy Narrows, Ontario, north of Kenora, for a number of years. There, a forest industry company was found to have polluted a river with mercury, which has injured many in the community because of their dependence on fish from the river as a major part of their diet. The issue has broadened beyond the mercury issue into one of industry's clear-cutting infringing on the community's rights to the land.

The following quote is from one of several articles addressing that issue that has been in the news for at least four decades and is still current:

> More than 100 supporters from across Canada and the United States are in Grassy Narrows this week for an Earth Justice gathering to raise awareness about indigenous rights and protection for the boreal forest.

> The event features a tour of a clear-cut area, sweat lodge ceremonies, traditional feast, training in non-violent protests and speeches by Grassy Narrows residents and other First Nations leaders who will reiterate calls for an end to the clear-cut logging on the band's traditional land-use area.[11]

For decades, non-governmental environmental organizations (NGOs) have been in conflict with the forest industry over some of the same issues that upset First Nations, and the following quote is an example:

> Armed with a new report, environmentalists have taken another shot at the forest industry practice of clear-cutting.
>
> The Ontario chapter of the Canadian Parks and Wilderness Society (CPAWS) has called on forest companies to adopt alternatives to clear-cutting to better protect the environment and ecosystems of the boreal forest.
>
> "Clearcutting harms wildlife habitat, water quality and biodiversity," report co-author Chris Henschel of the CPAWS-Wildlands League said Monday.[12]

It is not surprising that such tensions exist, because so many depend on the forest for so much. With an ever-increasing number of people but an ever-shrinking area of forest land available to them, conflicts inevitably arise. When humans first evolved, our first priority with the forest was to extract the raw materials for food, clothing, and shelter that were necessary for our survival. In those earliest days, there was no real conflict regarding harvesting food and wood from the forest because the number of people was small and the forests were so vast. As our populations increased and we spread around the world, however, our pressure on the forest increased. More and more, we used what tools were available, or we created, to extract from and change the forest to meet our increasing demands. The first tool was possibly fire and it appears to have been used by all humans, wherever we went, to change our surroundings to our benefit.[13] The clear-cutting tool was developed much later by an evolving human society.

We have since increased in sufficient numbers and developed the technology to eradicate all forests and change the world's entire ecosystem. Of course we didn't collectively intend doing such a crazy thing, so some nations have enacted laws to control their citizens who would, through disregard for others, endanger the well-being of humanity. Canada is one of those nations.

That's enough about economics and conflict. The focus of this book is on how forests can be best managed so that they will remain healthy and substantial enough so that all of the various interest groups can enjoy them. I want my readers to understand how the northwestern Ontario (NWO) boreal forest behaves when it is left to either, or both, natural and human influence.

## THE READ FROM HERE

In Chapter 2 I introduce myself and my early perceptions of the forest, foresters, and the forest industry. We then follow the trail that changed me from one who perceived foresters and the forest industry as abusers of the land to deciding to become a forester and work for the industry.

Chapter 3 opens with a short summary of my experiences as a professional forester and how those experiences impacted my earlier perceptions. My career took me from Newfoundland to Ontario, and from both locations I visited other forests across our nation. I became aware that the boreal forest is not a uniform, static entity, but a living, dynamic organism in tune with local climate and soil. Ontario's forest policy documents and their foundation are discussed from my perspective, which is based on my past involvement. I stress that more public and stakeholder knowledge about the forest, as well as a willingness to share, are needed as essential ingredients for future forest health.

Chapter 4, the longest, describes the overall structure of the NWO boreal forest and its renewal agents, followed by descriptions of the predominant tree species found in the boreal forest and how each interacts with its environment. Fire and its overbearing impact on the natural NWO forest takes up Chapter 5.

Chapter 6 gets into the crux of the book, and that is the role of timber harvesting in the NWO boreal forest, and an unabashed argument in favour of even-aged management and clear-cutting. That is followed by arguments against uneven-aged management and selection harvesting, noting their unsuitability for that same forest.

Chapter 7 describes some of the improvements that have been made in logging practices, and the strengths of current legislation, forestry funding methods, and independent forest audits. In Chapter 8, the concluding chapter, I invite stakeholders, environmental NGOs, and anyone else with a vested interest in the forest's health — that's all of us — to commit to finding common ground. If we work together on that common ground, we may be pleasantly surprised at the number of previously unforeseen opportunities that arise to work together for the good of the forest and ourselves.

# 2

## HOW I BECAME A FORESTER

The boreal forest has been a part of me, and me of it, from my birth. I grew up in a logging and wilderness-access village in the boreal forest of central Newfoundland, when what is now the newest province of Canada was a protectorate of the British Empire. My hometown of Millertown was the headquarters of a logging division of the Anglo (Nfld.) Development Company Ltd. (AND). The AND owned what once was one of the largest pulp and newsprint mills in the world, which was located some one hundred kilometres down the Exploits River from our village.

My father scaled (measured) wood, the basis for paying the company's loggers and maintaining inventory records. The company encouraged salaried employees to take their sons for visits to their bush camps. I recall spending some of the most enjoyable weeks of my life with Dad as he went about his job. On warm summer evenings, we sometimes fly-fished for brook trout for the camp's breakfast.

By the time I was ten years old, I was spending time alone in the bush behind our house. I recall that at age fourteen my best friend and I were wandering farther from home and exploring new territory. Whether alone or with friends, I was acutely aware of the forest as it was going about its business. Everything caught my attention, from the trees to the tiny mosses, from the moose to the weasels, voles, and green frogs, from the eagles to the chickadees. I was intrigued at how some species appeared to be dependent on some other species for their food and homes. I often sketched what I saw and collected books that helped me identify and learn about the different species. I recall that my habit of declaring the species of something caused my buddies to roll their eyes. They probably knew me as a nerd before Dr. Seuss had coined the word that fit me.

Millertown had an active Boy Scout troop and Girl Guide group. Our troop leaders were usually familiar with the local forest and living off the land and water, and they ensured that each scout developed those skills. Although I had a love of the land and enjoyed learning how to better interact with nature, I didn't harbour any ideas at the time of a career involving the land. Instead, I was developing a dream to become an airplane pilot. Millertown is located on the shores of Red Indian Lake, which is an ideal landing site for large amphibious aircraft. Although only five or six years old at the time, I can remember one day, during the later years of the Second World War, two Yankee Clippers landed and taxied toward each other. A group of American military personnel, who were stationed in an adjacent community, used a row boat to ferry people from one aircraft to the other.

Millertown was also beneath the flight path of the Ferry Command, which ferried military aircraft manufactured in the

United States and Canada to Britain. I recall hearing the distant drone of approaching aircraft and looking up as they passed in formation en route to Gander, where they refuelled. Following the war, bush planes began to appear in increasing numbers, and I recall a war veteran and bush pilot making numerous landings with his float-equipped Stinson on our lake. Every time a plane landed, I was at the lakeshore dreaming of the day I would pilot one like that.

## A FORESTRY SEED IS SOWN AND NURTURED

After finishing high school, I worked and lived in the bush camps and banked my earnings to finance my education but waffled on what I wanted to study.

One day a crew of forestry students led by an AND forester stopped at our camp. I spent several hours questioning them about their "forestry-survey" work and its purpose. After a while, the company forester, who had been listening to my questions, took me aside and asked if I might be interested in becoming a forester. Even though I had for years known company foresters, it was the first time I had ever thought about that possibility. Frankly, I did not approve of what the foresters that I knew were doing to my beloved bush.

As a bush worker, I was a member of the logger's union and during the winter of 1957 the International Woodworkers of America (IWA) began raiding us.[1] That February I was employed at loading pulpwood on tractor sleds and one evening, when we loggers returned to our bunkhouse, we discovered that it had been completely reorganized into a meeting room. After dinner, we returned to the bunkhouse where we were met by two strangers who told us we were having an important meeting and to take a seat. The meeting was called to order by a third stranger, who passionately boasted about what the IWA had done for loggers in British Columbia. As he elaborated about the pay and living conditions they had, he promised that, if we were to sign a card that we would be given, he and the IWA would ensure that we got the same benefits.

Then he asked, "Does anyone have a question?"

There was a long silence. He had described the menu offered in the B.C. logging camps and it was quite different from what we were eating, but in reality, our menu was much like what we were eating at home; except in the camps we were getting larger helpings and more desserts.

I was happy with our menu and, as a naive eighteen-year-old, had the bad sense to ask, "Why do you think I want something different from what I already like?"

From behind me, the two strangers who had ushered us into the bunkhouse took my arms and, lifting my puny, fifty-seven-kilogram body from my seat, led me through the door and told me, "Don't come back. You won't be welcome."

I retreated to the foreman's quarters where, after telling him of my plight, I was offered a bed for the night. On the hand-crank magneto telephone he called the divisional superintendent, who was apparently unaware of the IWA organizers' presence in his division. The next morning the superintendent showed up and offered me a non-union job in the depot supply store until spring.

Not long after my move, the superintendent introduced me to the company woods manager. The manager told me that he had heard about my predicament and had enquired about my employment record.

"You are wasting your time as a logger. Your record indicates to me that you can have a better future." Continuing, he said, "I guarantee you that as long as I am manager here you can have work, but I advise you to seek further education."

"I already have plans for that," I replied. "I am applying for entry at Memorial University of Newfoundland (MUN) for this coming fall."

After questioning me further and learning that I had no definite planned specialty, other than maybe a general science degree, he suggested, "You should consider studying forestry. If you do, I recommend that during your studies you look for summer work with other companies or governments to broaden your experience." Then he added, "Any summer that you can't find work, I will see that you get a job with AND."

With my love of the bush, the forester's encouragement, the students' inspiration from the previous summer, and the manager's promise, I made my decision. That winter I applied at MUN for entry into their pre-forestry diploma course. MUN had an arrangement with the University of New Brunswick (UNB), whereby an additional three years at the latter school could lead to a Bachelor of Science in Forestry (B.Sc.F.). I was accepted and registered that fall.

At the back of my mind, I still saw myself as an airplane pilot and had also applied for admission into the Royal Canadian Air Force (RCAF) Regular Officer Training Plan (ROTP). I had an ulterior motive; acceptance into the ROTP would guarantee that my education would be funded by them. At registration, I applied for the ROTP program and apparently qualified.

After a two-week cooling-off period, I was interviewed by a perceptive recruiting officer who quickly got me to admit that my main motive was to have ROTP pay for my education. He said, "Now look, Malcolm, you are committing yourself to five years in university. That's all to the good, but do you realize that upon graduation you are committing yourself to a five-year commission as an officer in the RCAF?"

"Yes, sir, I do," I replied.

"Well, okay," he continued. "What use do you think your forestry degree will be to you after that?"

I recall that he offered a few days for me to think on that and also assured me that, if I decided to withdraw from the program but changed my mind again, I could reapply the next year.

"That won't be necessary, sir," I replied, "I want to be a forester. You have helped me realize that and I thank you. I withdraw my application for admission into the ROTP program."

He stood up, shook my hand, and congratulated me on a good decision. From that moment on I have been committed to forestry.

## DECISION MADE — NOW FOR THE HARD WORK

During my student years, I followed the woods manager's advice and during summers worked with Fraser Papers in New Brunswick, Forestry Canada, and Parks Canada. For the final summer, I applied to AND, where the woods manager proved to be as good as his word. There, for part of the summer, I studied a new harvesting system that they were testing and used it as the subject of my senior report that was required for graduation.

All of my summer employers invited me to apply for permanent employment upon my graduation.

In the spring of 1963, despite additional interesting offers, I chose to return to Newfoundland and work for Price (Nfld.) Pulp & Paper Limited, formerly AND (in 1961 AND had been purchased by Price Brothers Limited of Quebec City). All through my youth, right up to graduation as a qualified forester, I was becoming increasingly contemptuous of clear-cutting and even-aged management in the boreal forest. With my forestry degree, I was now determined to return to my home and be instrumental in making changes that I felt were necessary to sustain the health of the forest.

# 3
## THE BOREAL FOREST NEEDS SOUND SCIENCE

I do my best to be objective. However, like you, I have experience-based biases, and you should know what mine are and how I acquired them.

I initially chose to work for the forest industry because I believed that industry was best positioned to advance forest management; all that was required were positive incentives, and strong legislation that was enforced on companies that failed to otherwise respond favourably. The Anglo (Nfld.) Development Company Ltd. (AND) had extensive private land, and on the Crown land it managed, it had ninety-nine-year renewable leases and licences. The timber leases and licences came with no stumpage fees (fees per volume of wood harvested); rather, they were associated with area-based, land-rental fees. The combination of secure long-term tenure and area-based fees prompted them to search for ways to maximize yield from the land.

At Price (Nfld.) Pulp & Paper Limited, I came under the direction of a highly respected forester, Frank R. Hayward, who took an intense interest in my training. He was open to hearing my perceptions of what good forest management looked like and at no time told me I was wrong about anything. He was, however, skilled at getting me to question my objectivity.

He was unfazed by my assertion that clear-cutting was wrong and didn't ask me to explain why I believed that was so. The fact was, from my earliest childhood experiences on logging operations with Dad, the sight of dense piles of drying treetops, dead and dying herbs and mosses on the extensive cut areas revolted me. I thought of all the wildlife that, as a result, were displaced and the baby birds that probably died when the trees were harvested.

It never occurred to me that the forest fires that frequently burned in our area did the same thing, and more, by often killing all life above ground in their paths. All through forestry training, despite what I was learning in silviculture and forest management courses, and even through my first four to five years of forestry work, I continued to be revolted at clear-cuts.

I hadn't been working under Frank long before he outlined what he saw me doing over the next several years. He told me about the company's long history of harvesting and stand-management research and suggested I spend some time getting familiar with the records. In their fire-proof vault, I discovered file cabinets full of field data and interim reports. Some of the data and reports went back as far as 1921 and had been authored by John D. Gilmore. John was an early graduate of the University of Toronto's Forestry School and had joined AND in 1918, later becoming woods manager and starting the various "trials."

The trials covered a variety of stand-management methods and logged-area treatments, including different prescribed burning techniques over different establishment years. Logging slash (discarded branches and tops of harvested trees) treatments and prescribed burning of harvested areas were company practices through the 1920s.

The records revealed to me that in its early years AND had been operated according to the values of its owners who were familiar with current European forest management methods. They wanted to ensure that their harvesting encouraged healthy regrowth and they intended to periodically thin the new stands as they developed. The company had for decades been trying to determine the best silvicultural (forest farming) practices for the forest under its control.

That effort continued following the Second World War, when AND in Grand Falls, Newfoundland, and Bowater Pulp & Paper Ltd. (Bowater) in Corner Brook, Newfoundland (the original Bowater paper mill and at that time also British-owned), jointly formed the Anglo-Bowater Forestry Research Organization. In 1946 they hired W.M. Robertson, the newly retired director of the Canadian Forestry Service, to manage the organization. He added additional stand-management trials across AND's land holdings and established 999 permanent growth-and-yield sample plots (PSPs), which were measured four times over three decades.

Scheduled re-measurements weren't due for any of the trials for another three to five years when I joined the company, so I had time to see them in the field and plan my schedule. Over a two-year period, I cruised timber (collected tree and stand data), surveyed and mapped stand depletions, and completed a survey of tree regeneration on over five hundred square kilometres of former cutovers in the Noel Paul River watershed. The survey showed that the cutovers had regenerated to the species harvested, but moose, which had been introduced, were eliminating balsam fir, birch, aspen, ground hemlock, and most hardwood shrubs from approximately 20 percent of the watershed and the most fertile soils. Most of those areas were converting to grassland with scattered white spruce.[1]

I was now even more skeptical of clear-cutting. After two years with Price (Nfld.) Pulp & Paper Limited I was becoming disillusioned and skeptical of the possibility of the chances of my goal to make positive changes and began looking for other opportunities. I resigned from Price (Nfld.) and was hired by Forestry Canada as a forest research officer. With them I hoped to study tree and stand growth relative to tree location and spacing within a stand and various stand-management techniques to vary that spacing and improve growth. I quickly began to feel that I had made a mistake, however, as I became disappointed and impatient with the slow pace and what I felt were inefficient civil service policies and practices.

At this time, I had, with Forestry Canada's encouragement, applied to a number of Canadian and American universities for post-graduate studies, and Yale University chose to accept my application and offer me financial assistance. That was during the Vietnam War and I was advised to register for the military draft immediately upon arriving in New Haven. Enquiry showed that as a Canadian citizen residing in the United States I would not be drafted if my "number came up." However, America was going through a period of intense self-examination and I feared that the military draft could be broadened. I didn't like even a slim possibility of being drafted, and that, combined with my disappointment with the slow progress of my research proposal through the civil service hierarchy, and limited ability to access the bush except in summer, was enough for me to consider other options.

Price (Nfld.) Pulp & Paper Limited had been giving signals that they wanted me back to direct their forestry program and to serve as an understudy to chief forester, Frank Hayward, who was due to retire. The offer was certainly an improvement over the situation I had left when I was last with them. I felt that if I were to become chief forester, I would be better able to implement some of the forest management change that I felt at the time was essential for forest health. I took their offer and remained there for twelve enjoyable years, directing their forest management and forestry research.

During that period, I oversaw the re-measurement of all of the long-term trials and wrote, or supervised the writing of, final reports. Some draft reports were reviewed by selected external experts and they were revised accordingly, but none were subjected to the rigour of official peer review and publication in accredited scientific journals.

Based on the results of trials initiated in 1921, which showed significant merchantable-yield improvement,[2] we initiated several small pre-commercial thinnings of dense young stands. The thinning removed excess tree saplings, while selected desirable saplings were left standing at as near as practical to two-metre spacing. The thinning recovered no merchantable wood and therefore yielded no immediate cash return. That type of thinning is today called "spacing," or "cleaning." In 1970, after improving labour productivity and thus the cost of the thinning, we began expanding the program, with both federal and provincial help, to over a thousand hectares per year.

As part of my responsibilities, and as an interested tourist, I travelled widely in Canada, parts of the United States, and New Zealand, familiarizing myself with other companies' and various governments' policies and practices. I visited the forestry operations of numerous companies in various provinces and states, including, among others, the clear-cut mountain terrain of MacMillan Bloedel's business in the Vancouver Island rain forests; Proctor & Gamble's operations at Grande Prairie, Alberta, and its clear-cutting in the lodgepole pine forest of the Rockies' foothills; Weyerhaeuser's Douglas fir plantations near Mount St. Helen's in Washington State; California's redwoods; New Zealand Forest Products (NZFP)'s intensively managed radiata pine plantations on North Island, New Zealand; J.D. Irving's operations in New Brunswick, with its extensive black spruce plantations situated on former clear-cuts that had been prepared for planting with heavy vegetation crushers; and Spruce Falls Power and Paper's fill-in plantations on the clay belt forests of northeastern Ontario.

In almost every case, to the best of my knowledge, the forests were being managed according to the current laws of each jurisdiction that, compared to today's laws, were rather lax. Except for Spruce Falls Power and Paper, companies that were practising the more intensive forest management, at that time, put their greatest effort into their private land.

Throughout my career, I was sometimes assigned to lead bush tours for Canadian and foreign politicians, stock analysts, buyers, and executives of international newspapers who were our current or potential customers. Conversations during those tours exposed me to the politics of world trade and investment and gave me some insight into the complexity of satisfying customer demands. To my surprise, puzzlement, and gratitude, one newspaper executive wrote our CEO giving me credit for his paper's large newsprint order.

In 1974, Abitibi Paper Company acquired Price Brothers Limited, and the combined companies were later renamed Abitibi-Price Inc. Philip Mathias's book, *Takeover,* provides a detailed and intriguing replay of events leading up to and during the high-stakes boardroom, stock market, and financial manoeuvring that occurred at the time.[3] In 1978, I was temporarily transferred to head office to work with Duncan Naysmith and Frank Robinson of Abitibi-Price as they negotiated Forest Management Agreements (FMA), first with Manitoba and then with Ontario. The first FMA negotiated in Ontario, for Abitibi's Iroquois Falls Division, became the trend-setter for two additional agreements for Abitibi, and subsequent FMAs that followed with all other companies in Ontario. FMAs transferred responsibility for stand renewal after disturbance and follow-up stand management from the Ministry of Natural Resources and Forestry (MNRF) to the industry. That exhilarating experience culminated with Bill Johnston, the vice president of woodlands for the company, challenging me, "You have helped negotiate FMAs, now make one work."

I was made accountable for forest management of the Spruce River FMA, miscellaneous licences, and Abitibi-Price's developing forestry program on its two thousand square kilometres of private land northwest of Thunder Bay, Ontario.

With the exception of approximately one year working for Forestry Canada and a summer job with Parks Canada, all of my employment between 1956 and 1997 was with Canadian forest-industry companies. I left Abitibi-Price in 1997 to become an independent forestry consultant. For the next eight years, under short-term contracts, I supervised forest renewal, advised on forest silvicultural and urban forestry projects, was lead auditor on two independent forest-management audits,[4] and assisted the successful bidder when what had since become Abitibi-Consolidated sold its private land.

During my career, various persons provocatively challenged me with, "How can you, a professional forester, promote industrial forestry and not be ashamed?"

My answer was always the same. "I am proud of what I am doing with the forest industry." And then I'd proceed to explain why to anyone willing to listen. The remainder of this book is my expanded answer to the question.

## THE FOREST IS MORE THAN ITS TREES

As a teenager, I often travelled along the numerous fens, muskegs, and barrens of my Newfoundland home, where less than 50 percent of the landscape is closed forest. The fens at valley bottoms, muskegs, and rock barrens on the ridge tops and some slopes, support few trees.

During my work there, I used those open spaces as travel corridors, particularly in winter when they offered unimpeded snowshoe, ski, and snowmobile travel on the wind-packed snow. I would stand gazing up from the side of a fen or down from a ridge top at the opposite hillside, studying the layout of the tree cover, its species content, and relative tree sizes and stand volumes.

Over the years, those images converged in my mind with numeric cruise data collected along parallel lines through the forest and I could give increasingly accurate ocular estimates for portions of a stand. By examining aerial photographs, I could then extend the estimate to a whole watershed. I was unaware that a mix of forest and open spaces was being burned forever into my psyche.

When I was transferred at middle age to Ontario, I initially felt uncomfortable in the boreal forest of this new place, and that bothered and puzzled me. More than ever, I was enjoying being on large clear-cuts, and one day, during my second year in my new job, while kneeling and examining some newly planted spruce seedlings, a sudden realization came that I was missing the open spaces that I so enjoyed.

I, a forester, was embarrassingly claustrophobic in Ontario's relatively unbroken forest, where even the fens and some of the muskegs tend to be treed. I realized that the clear-cut and burned areas had become surrogate fens, muskegs, and rock barrens, and views from them were now adding to my understanding of the local standing forest. I was also appreciating clear-cuts for their open space and I could now even see their beauty.

Clear-cuts and burns, which in my youth had offended me as eyesores, had evolved in my mind into places of wonder and what was initially stark beauty, into increasingly appealing panoramic scenes and smaller scenes of vivid colour and form.

Clear-cuts and burns became subjects of my expanding visual-arts hobby and small business. I admit that it takes a stretch to see beauty during the first couple of years after any forest disturbance, but by the third year, the new growth of the remaining trees is accompanied by flowering herbs and shrubs, and a wider variety of mushrooms begins to appear.

As a student, I worked one summer with a scientist who developed one of Canada's earliest ecological forest-site classifications, and with him examined several hundred small sample plots of the forest floor.[5] Throughout my career, I used similar plots from which to gather information that enabled objective analysis of various forest dynamics. Forest ecology didn't become my life vocation but I did learn that the forest is more than just its trees.

The number of species of trees in a stand is a small fraction of the total number of plant species. During much of my time in the forest, I have been on my knees and even lying face-down while examining the diversity of life on a few square centimetres of forest floor. Photographing or drawing, collecting, and identifying as many plant species as I could find was an early hobby, and as a forester I often recorded and collected for my learning and research, and research by others.

I have accumulated a large library of photographs of fascinating rock patterns and colours, lichens on rocks and wood, mosses, ferns and grasses, dew-covered spider webs, intriguing arrangements of rotting wood, bark, and fungus, and, yes, wildlife, including insects, birds, frogs, snakes, turtles, hares, grouse, porcupines, foxes, lynx, wolves, coyotes, bears, moose, and caribou — practically all in clear-cuts and burns of differing ages. Yes, beauty can be found anywhere but it takes the right attitude and focus on the here and now — and it helps to forget our biases.

Plants are the base of all other boreal forest life, providing food and shelter to wildlife, and food, medicine, tools, shelter material, industrial raw material, and spiritual comfort to humans. Additionally, each species performs various functions within its immediate environment as it integrates and competes with its neighbours. Simply put, without plants we can't continue to exist but they can do quite well without us. Provided we use plants responsibly and work with natural processes when harvesting them, they will continue to sustain our species and our way of life.

## WE ARE LIVING IN AN EVER-CHANGING FOREST

I hope to stimulate examination of the forest around us and objective questioning of much of what we currently hold as fact. Please critically question everything that I and others are telling you. I hope that you will seek other sources of information and weigh the "facts" as presented by each of us. We all see the world through the clouded lens of our own experience and too often assume we know best. The deteriorating condition of our world is proof that approach has not been good enough.

The internet has numerous scientific articles that are easily accessed. Do an online search for "black spruce" to see what information is available about that species. You will discover in your reading that our knowledge of trees and the forest is limited, but as with other subjects, we are rapidly adding to what we already understand, and in some cases altering and even rejecting what we previously thought to be fact. That is how objective science works.

We don't know everything about the human body, even though for centuries it has been scientifically studied. Consider all of the forest-tree, other-plant, animal, and insect species that together have received only a fraction as much scientific examination and you will understand the vast amount of knowledge that is waiting to be discovered about the boreal forest.

We know that the boreal forest is constantly changing and that sometimes alteration is slow and unnoticed by the casual observer. At other times change is rapid, with hundreds of hectares harvested or even a thousand square kilometres transformed from mature forest into a large, wildfire-blackened landscape.

If you own a summer cottage in the bush, think back to when you first saw it or cleared the site for building. Check the photographs that you have taken over the years and observe the change in the forest. I expect it will surprise you that new trees have appeared, some have disappeared, and most have grown without your notice.

Some of us return again and again to our favourite locations in the forest and they always appear the same, but are they? Think of your favourite blueberry patch. It wasn't always such. Probably only eight to ten years ago it was a mature forest that was burned or harvested. Two to three years after the disturbance, the blueberry plants matured, and since then crops have been variable, with one or two bumper years. Then, last summer, you realized that the developing young trees are shading out the blueberry bushes. Next year, it will be time for you to find another location.

Between high school and forestry graduation, I gained experience at most jobs involved with circa 1950s' pulpwood harvesting and extraction. I was interested, but skeptical, when some of the oldest loggers said that they were harvesting the same ground for their second time.

In 1949, while accompanying my father as he scaled stump-piled pulpwood (wood cut up and piled where the trees were felled), I was overwhelmed by black flies. He sat me on top of a pile of logs where a breeze kept the pests at bay while he carried out his work within sight of me. In 2000 I returned to the same location with my adult son. The site had again been recently clear-cut and my son obliged me by standing beside a pile of logs near where I had sat on another pile fifty-one years earlier.

As a logger for a brief period in 1958, I helped clear-cut a portion of a stand that had been harvested circa 1908. During a visit to the same location in 2011, I was pleased, but no longer surprised, to see that it was ready to be harvested a third time at high commercial volume — and that on Newfoundland's supposedly poor soils.

Since moving to Ontario, I have examined ground on the former Abitibi-Price's private and public licensed land, where second clear-cut harvests have occurred. In 2011, I returned to a 1983 black spruce plantation on a former clear-cut within the Spruce River FMA. I was so impressed that I surveyed a part of the stand and determined that at only twenty-eight years since planting it already had as much pulpwood volume per hectare as the average tract that Abitibi-Price had harvested in its early clear-cuts.[6] For black spruce that is remarkable, especially when one realizes that we have traditionally regarded that species as taking one hundred years to mature in Ontario.

That particular stand was the subject of considerable public and professional criticism as we began taking responsibility for stand renewal. Many viewed our efforts as failures from the start and our credibility was on the line. Then and now, we all had and still have a lot to learn!

A bright and satisfying future awaits anyone interested in studying trees, their growth, their interaction with the environment, and their chemical and physical structures. Lakehead University, in Thunder Bay, and several other universities across Canada offer tremendous opportunity to anyone with the academic credentials for and an interest in such studies.

## FORESTRY PRACTICES GOVERNED BY LAW

From the early 1960s, Canadian citizens began to take a more active interest in the way we manage our forests. It was discovered that not all was as we wanted and we became more proactive.

As a practising industrial forester, I took the brunt of many citizens' disgust with what was perceived as the irresponsible behaviour the forestry industry had displayed in its forest-management practices. Gradually, through the 1970s and 1980s, that disgust fed the creation of powerful citizen lobbies led by passionate activists and it found its way into our schools and other public institutions.

I remember loggers emotionally appealing to me to do something on their behalf to "get the truth out there." Their children were being bullied by their classmates and many were ashamed of their parents' jobs. Those were frustrating, maddening times, and it was difficult for forest workers to keep their cool and remain objective.

Until recently, industrial foresters have had an almost impossible job getting our perspective into the public arena. Those of us who attempted to get our story out were often not skilled at working with the media and consequently often looked incompetent, antagonistic, and even deceitful. Let me try to explain how things stand today.

Since 2010, some of the best known and progressive environmental action and lobby groups have taken a different approach. Groups such as the David Suzuki Foundation, the Canadian Boreal Initiative, Forest Ethics, and the Canadian Parks and Wilderness Society (CPAWS), among others, decided to look for common ground and work with the forest industry on agreed objectives. Together they have formed the Canadian Boreal Forest Agreement, which currently includes a total of six environmental groups and eighteen forest industry companies as members.[7]

Greenpeace remains a notable exception and has chosen to continue its confrontational methods. Significantly absent as partners to the agreement are the First Nations, who say their treaty rights are being ignored.[8] Hopefully, in the not too distant future, all parties will come together to work on the common cause of sustaining the boreal forest. It's time to face up to the reality that there will never be total agreement on all issues, and the best way to realize improvement is to work together on those issues that we can agree on.

Forest management in Ontario and across Canada is guided by policy developed under the collective wishes of citizens expressed through our elected representatives and the legislation and regulations passed by provincial parliaments. All foresters practising in Ontario are accountable for their professional actions.

When I arrived in Ontario, my employer required that I join the Ontario Professional Foresters Association (OPFA) and become a registered professional forester (RPF). In order to become a member, I had to pass an exam as evidence of my knowledge of past and current Ontario forest policy.

The OPFA was created by the Professional Foresters Association Act on April 3, 1957.[9] That act, requested by a core of dedicated foresters, gave them a rallying base from which to collectively increase their forestry competency and standards of practice. Since October 16, 2000, the practice of forestry has been subject to the Professional Foresters Act, 2000, which recognizes professional forestry as an independent self-regulating profession.[10] All RPFs are governed by the requirements of that act and our code of ethics that can be accessed through an online search of the OPFA website.

Now that I am retired and inactive, I have chosen to focus on the concluding words of "A Commitment to Learning," the last value in that code, which states, "A member shall ... use their knowledge and skills to aid public awareness of forestry in Ontario."[11] I am doing that through direct communication via one-on-one conversations, public speaking, leading forestry tours, and with my visual art and writing.

The forests that exist today in Ontario are a reflection of past citizens' demands (or lack of) on politicians, the politicians who were elected, their willingness to enforce legislation once enacted, and, to a large extent, natural events that were beyond human control. As proactive citizens in a democracy, such as Canada, we have the power to require our politicians to pass the legislation we want and to hold them accountable for implementing that legislation. However, the onus is on us to be well informed of the consequences of filling our demands because many of the demands we are making will not give us the forest we want or will need in the future.

## JUST HOW DYNAMIC IS THE BOREAL FOREST?

No individual or group has complete knowledge of the forest, so sharing what knowledge we have is essential to achieving responsible management. Sitting as we "northerners" do in the middle of the boreal forest, we have a responsibility to ensure its continuing health.

I spent most of my boyhood roaming the bush where I developed a life-long love for things natural — the scenery, rocks, water, flora, and fauna, the whole kit and caboodle. That doesn't mean that I was a tree hugger or an animal-rights activist, no, not at all. Instead, I made use of what I loved because at an early age I realized that I was a legitimate part of the natural world in which my species, like all others, had evolved through competition for living space and food.

As I observed the relationships among the different parts of the forest, I discovered that each individual living part of it was dependent on utilizing some other part, sometimes to the benefit of the one being utilized, but sometimes not. Sometimes the use of a species by another affected a third species — sometimes negatively and sometimes positively. An example of the latter scenario would be how dragonflies prey on other insects, such as black flies, deerflies, and mosquitoes, which in turn prey on moose and humans. The moose and we benefit from the dragonfly's predation.

Through my observations, I also learned that there is no single "balance of nature." Nature is in constant flux. What today appears to be in balance may tomorrow be thrown off balance by some intervention, be it from another species, including man, or some weather, atmospheric or, potentially, celestial event. No patch of forest that one sees today will ever again be exactly the same, because as we observe the woods some plants are growing while others are declining and, with that change, mammal, bird, and insect populations also modify. Even the soil may be changing. The interaction of so many variables becomes so complex that the chances of a repeat occurrence are infinitesimal.

After moving to Ontario, I familiarized myself and fell in love with this area of the boreal forest. As a practising forester in Newfoundland and Ontario, I prescribed clear-cut timber harvesting. Since my youthful opposition to clear-cutting, my training and personal observations have taught me that, if we are to utilize the trees of the boreal forest, and return similar species without too dramatic a change occurring, even-aged forest management is essential. Boreal tree species have evolved for millennia, accommodating disturbances by insects, disease, wind, fire, and, yes, man. This has resulted in a forest that is made up of large areas that contain trees that are all the same age. Adjacent areas similarly have trees of a single age — the age being dependent on the time of disturbance from which that area of the forest originated. We call that an even-aged forest.

Here in northwestern Ontario, forms of harvesting other than clear-cutting will more likely result in "uneven-aged" (multi-aged) stands, with a different mixture of tree species. For example, we will probably witness a transition from the current preponderance of jack pine and black spruce to more balsam fir and white spruce. What will then happen to any other plant or animal species that may rely on even-aged stands of jack pine and black spruce? I offer a partial answer to that question in the following chapters of this book. Please read on with an open mind.

## SUSTAIN FORESTS WITH WANTS IN MIND, BUT BE WILLING TO LISTEN TO REASON

Humans have been a part of the fauna of the North American boreal forest since what may have been the first group of immigrants to the Americas crossed "Beringia" (the Siberia-Alaska land bridge) during the most recent ice age. While living here, we have extensively changed the forest structure that will prevail well into the future.

Some believe that the boreal forest has increased in area since we stopped burning the prairie grasslands. Frequent fires are believed to have maintained the grass cover by burning trees that advanced into the prairie. Areas in the vicinity of the voyageurs' Dog River/Prairie Portage north of Thunder Bay, Ontario, are today forested with tree stands of various ages. However, after repeated fire or timber harvesting activities, tall grass quickly becomes the predominant vegetation at some locations as far north as Graham. As recently as 1856, approximately sixteen square kilometres of grassland survived near what is now the community of Stanley west of Thunder Bay.[12]

A century ago, because of catastrophic losses of life and property, we began controlling fire in the forest. As we improved our success at fire control, the forest gradually began changing — with light-demanding species being replaced by shade-tolerant species, and associated wildlife. The change was enhanced by expanding timber harvesting that used horses, particularly in winter, which caused little soil disturbance, enabling the shade-tolerant species, such as balsam fir and white spruce, to thrive.

A few decades ago, we got serious about managing our forest and began attempting to sustain former "natural" cycles and species. More sophisticated legislation and regulations passed responsibility to natural resource managers, primarily foresters, to manage the forest according to science-based guidelines while, where possible, accommodating changing diverse and conflicting public desires.

Our limited individual experiences and knowledge ensure that we tend to see our personal wants as being most important. How often do we ask ourselves, "If I get what I want, who must sacrifice what they want?" Sadly, some of us don't even care what the other person wants or needs. Natural resource managers can act only after weighing the perceived benefits of those actions against the impact they will have on what others want or need, and, of course, ultimately on the impact that they will have on the forest. Inevitably, some people are never pleased with whatever decision they make.

Today we hike, pick berries and mushrooms, fish, hunt, trap, sight-see, photograph, paint, canoe, camp, cut Christmas trees, mine, log, and carry on many other activities. Disgustingly, some of us carelessly dump our household and even small-business garbage in the forest. We live and move about in the forest in numbers never seen before. It behoves us to consider what kind of future forest we want.

Our governments have responded primarily to public feedback that has indicated that we want our future forest to be more like what existed before we began changing it dramatically. I believe that government has developed the appropriate legislation that can make that happen.[13]

Does that mean that fires should be more frequent and larger? If they are, life and property loss will possibly be greater than ever because of our expanded presence in the forest. I don't think that we should take that risk.

In my opinion, if we are to continue living in the boreal forest and sustain the forest so that, in total, it retains the diversity and scale that nature has provided, we need to control wildfire. However, if we decide to control wildfires, then at the same time we should continue timber harvesting in patterns that resemble those of wildfire. With known silvicultural techniques and the skills that we have today, we should regenerate the forest to fire-encouraged, light-loving species while, where reasonable, accommodating local individual wants.

## THE SHAPE OF THE FOREST

After the Second World War, we had a good idea of what a bird's-eye view of our forests was like, as by then black-and-white aerial photography was in common use. The company that I began working for after my graduation in 1963 had, in 1946, spliced individual black-and-white aerial photo images together to form mosaics of its entire licensed and owned areas. Each mosaic covered approximately one hundred square kilometres — quite a lot of ground. When several mosaics were joined, the black-and-white image didn't provide enough contrast and the large scale of one inch per one thousand feet (one centimetre per one hundred and twenty metres) made it difficult to view several large burns at once.

That changed in the 1980s, when small-scale satellite imagery became available. Land managers could then obtain colour-enhanced images that demonstrated the extent of, and relationship among, the different disturbances that impact the forest. It became clear that beyond the areas where our values, activities, and development had encroached, individual fires extended over hundreds and even thousands of square kilometres and were the dominant force in shaping the forest.

The historic pattern was of large burned areas that resembled giant jigsaw-puzzle pieces, which tended to be elongated southwest to northeast with fingers poking into stands that had originated following earlier burns. Within the large older burns were islands containing stands of different ages. Some of the islands were small, more-recent burns and some were older stands that had temporarily avoided fire because of variation in wind, terrain, underlying soils, stand content and structure, and other factors at the time of the surrounding fire. "Dirt" foresters, those with soil beneath their fingernails, had known those facts for decades, but with satellite, and later digital imagery and other advanced technology, the pattern was now available to see and manipulate.

Timber harvesting early in my career was limited by natural stand boundaries and the merchantability of stands and individual trees. Fire-origin stands were accessed at maturity and harvested over time, leaving unmerchantable portions standing. The resulting pattern of the harvested area thus tended to resemble that of the burn.

As concern about the sustainability of our nation's forests intensified through the 1980s, things like clear-cuts, roads, dams, open-pit mines, and just about anything that we deemed to be too large was described as "so large that it is visible from space." Such a description was not very meaningful since something doesn't have to have a very large area to be visible from space if one uses the right optical lens. For me, statements like that have helped damage the credibility of some environmental activists.

Nevertheless, public concern about the state of the forests did become a political issue. Various jurisdictions responded with rules and regulations ostensibly designed to help protect the forests from excessive commercial exploitation and environmental degradation. The Ontario government responded to public concern by scientifically studying the forest pattern and requiring intensive public hearings on forest management. Provincial governments of all political stripes used the findings to develop forest-management legislation, regulations, and guides requiring forest managers to design timber harvesting and renewal practices to imitate natural forest-stand patterns. To me, the harvested areas resulting from the new regulations resemble stand patterns of timber harvesting that prevailed during my early career. At that time, we harvested all merchantable species and tree sizes as we came to them, until we were blocked by unmerchantable portions or difficult terrain. The resulting cutover contained small patches of unmerchantable species and tree sizes, internally inaccessible terrain, and irregular external boundaries — the overall appearance resembled wildfire patterns. That is the pattern we are trying to create today as we follow current forest management guidelines. As described in Chapter 1, the more progressive nations that contain boreal forest are changing their forest management to better fit local historic natural forest patterns.

Like today, harvesting during my early career was sensitive to society's values of the time, limited by knowledge and available technology. I find it ironic that we appear to have come full circle in harvest patterns; earlier harvesting was left mainly to the harvesters, who followed practical patterns they found in the woods, whereas today foresters are following elaborate, digitally modelled, information-rich, multi-stakeholder driven plans.

Meanwhile, timber harvesting came through a period of intensive technological development that enabled harvesters, to some degree, to ignore stand and terrain parameters that fifty years ago prohibited economic harvesting. The result was a lessening of harvesting sensitivity to both our needs and nature's requirements, which in turn fuelled sustained negative public reaction.

Society has let it be known it has a voice, and in our democracy that voice will prevail. That makes it critical to forest health that our collective demands on forest managers be based on sound science and not on erroneous perceptions and narrow-interest wants.

## KNOWLEDGE PROMOTES RESPECT

Anyone who has travelled across Ontario and wandered in the boreal forest realizes that conditions on the ground vary widely across the north. In much of the northeast, glacial-lake deposited clay soils predominate, with poor drainage and resulting deep peats. Further west, rock of the Canadian Shield is virtually at the surface over extensive areas, and variable but generally coarser debris (till) that was left after the latest glacial period predominates. Over extensive areas, folded rock oriented southwest to northeast reveals bare rock through shallow soil on top of pockets of glacial till and waterlogged peat in the bottoms between the folds. The more porous till and underlying rock patterns determine drainage and, both directly and indirectly, tree-stand patterns.

Field-practising dirt foresters used to need several years in a rather limited area interacting with the local climate, soils, terrain, and biodiversity before becoming competent at interpreting the future results of forestry practices. A forester with several years' experience successfully regenerating stands on harvested areas in the northeast could initially experience difficulty interpreting stand succession and probable development after harvesting or other treatment in the northwest.

The MNRF recognized a need for tools to enable foresters to readily understand the unique nature of local forest conditions, and as a result it developed different ecological forest-site classifications and workable management strategies per site for the two regions. The ministry has published detailed guides that describe the various sites, their associated vegetation groups, plant succession paths, and workable soil and stand treatments.[14]

When I was amassing my experience, it took me several years at each posting to accumulate a level of competence that still had its gaps. Currently in Ontario, it is possible to shorten the learning period and also fill in those gaps. A forester who is already knowledgeable about soils and boreal-tree behaviour can follow the guides and, with some practice, competently prescribe treatments and anticipate their long-term results.

The guides are some of the many positive outcomes from Ontario's Class Environmental Assessment on Timber Management Planning (CEA). Prior to the Assessment Board's approval of the CEA and completion of its report in 1994, there was widespread public doubt about almost every aspect of forest management. Over a four-year period of intense quasi-courtroom hearings, backed by field examination, the Assessment Board examined every aspect of timber management as it was being practised. It heard evidence from a wide range of scientists and advocates and cross-examined witnesses from government, industry, lobby groups, and citizen coalitions. Where deficiencies were found the Board ordered a schedule of actions that included studies to improve knowledge, procedures to ensure continuous improvement, involvement of citizens, and stepped-up development of a variety of guidelines for planning and implementation.[15]

Of course the CEA didn't end all disagreements about timber or forest management, because we still have our individual disagreements with things that aren't going our way, and we continue to support a variety of lobby groups. Fortunately, the CEA supported the MNRF's involvement of Local Citizens' Committees (LCCs) to work with timber licensees and the MNRF during development and implementation of what are now Forest Management Plans.

In my opinion, the LCCs have been essential to responsible forest management since the early 1990s, and they continue to serve a valuable role in helping the MNRF, industry, other stakeholders, and citizens avoid destructive polarization, and merge conflicting ideals into workable plans.

Over the years, the LCCs have been served by a large number of dedicated volunteers. When members are exposed to the complexities of interpreting forest sites and prescribing management strategies to protect the environment, supply raw materials for society's economic and social needs, and satisfy individual wants, they become more willing to compromise. During my involvement with LCCs fifteen to twenty years ago, I was impressed with how industry, government, and citizens varied their wants and agreed on actions that they believed would be most beneficial to all. In my experience with LCCs mutual respect and compromise trumped polarization and conflict every time. That works in forest management and I believe that it can work in all of our endeavours.

Most of the individuals who joined an LCC did so because they, or their group, had a specific concern about the potential impact of forest management practices in a limited location, and others about the wide-scale impact. Cottagers didn't want to one day discover that forest around their properties was being harvested.

Mr. David Jones, a current LCC member and former school teacher, sums up his lengthy voluntary involvement as follows:

> My involvement with the LCC for the Black Spruce Forest started in 1992 when I was appointed to represent a group of recreational cottagers. Growing up in the Island of Borneo and my participation in summer employment with MNR contributed to my keen interest in forestry. As committee members, we have a genuine concern for the environment and a desire to make a contribution to forest planning. Each of us brings a diversity of perspectives and multiple points of view into the forest management process. I feel our contributions are valued and effective at influencing local forest decision-making. I continue to value my participation and to affect change in forest management practices.[16]

Mr. Maurice Rubenick, another member of the same LCC, comments as follows:

> [In] the middle '50s, [when I was] a young boy, I found freedom, adventure, and peace of

mind [in] the forest environment. In the '70s, if I came to places on Crown land where the MNR signs would say no motorized vehicles past this point, it really bothered me, because it's Crown land. A job-related accident in 1988 put me in a wheelchair for the rest of my life. I was depressed but still had an urge to be at one with nature. So, when a call went out in the early '90s for persons to join the LCC, I thought that maybe me and my wheelchair could make Crown land more accessible for seniors and persons with disabilities.

I believe that what I bring to the committee is a different look at forest management and how it's done. For the betterment of everyone. Just the appearance of my wheelchair at committee meetings makes them all look at issues a bit differently.

And now, after twenty-plus years, the diversity of the membership has made outcomes on issues much better for all. I also believe that with the diverse LCC there the companies and MNRF find it easier to deal with issues (their thoughts on issues are scrubbed and washed at committee meetings and they turn out proper for all in the end).

We've come a long way with this LCC, and if you were a fly on the wall at one of our meetings, you'd see the whole group looking at an issue from all different angles with open minds.[17]

Participants, such as Mr. Jones and Mr. Rubenick, have helped avoid most potential conflicts among forest users. Each member represents the vested interests of groups, such as, in Mr. Jones's case, cottagers, and in Mr. Rubenick's case, anyone wanting access to otherwise denied locations because of personal handicaps. Others represent First Nations, trappers, hunters, fishers, and berry pickers, to name a few of the interested groups. In all cases, these groups exhibit a general concern for the forest and human health. The Black Spruce Forest LCC currently has eighteen members, two representing the general public, two First Nations communities, and fourteen representing other organizations and vested parties. One additional member represents the prime licensee, and three others bring the interests of the MNRF to the group. Through discussion at their regular meetings, each learns from the others as they try to resolve conflicts and bring as near as possible a consensus to the planning process. After plan approval, their interest transitions to helping to ensure the forest licensee complies with the plan, or, where and when the unforeseen occurs, to helping to ensure that the plan gets amended.

# 4

## EACH SPECIES HAS
## UNIQUE REQUIREMENTS

**Y**ou now know that I am aware of at least some parts of the forest other than its trees and by now you have detected some of my biases and learned how I acquired them. You are also aware of how I acquired my limited knowledge about the forest and to some extent human nature. Now let's get down to the nitty-gritty of my understanding of how the boreal forest works.

We need to start by acknowledging the overbearing impact of fire and examining the needs of the individual tree species that inhabit the boreal forest of northwestern Ontario (NWO). Fire has occurred as a semi-random event in the boreal forest since the forest migrated north after the most recent ice age. All of the boreal forest has burned probably at least a few times and much of it possibly as many as a hundred times. Each fire has a different impact on the future forest, depending on various factors, including the pre-existing forest and the types of soil and terrain it covered, the season of the fire, and prevailing weather. For those reasons, foresters often describe the boreal forest as "fire driven."

Although there is certainly a degree of variability in the makeup of the forest as it regenerates after a fire, some results tend to repeat on burned ground in NWO. Jack pine, black spruce, trembling aspen, and white birch trees are common to almost all burned areas because of the unique ways each species has evolved to fill fire-created niches.

The portion of the forest dominated by each species is determined by the needs of each species, the particular mix of factors just mentioned above, and many more determinants. What needs to be understood, however, is that these trees are not randomly scattered throughout the forest. Rather, they tend to occur singly or as different mixtures and in groups throughout theoretically uniform stands.

Foresters are often condemned by well-meaning critics for creating monoculture tree stands that are vulnerable to pests, but single-species tree stands are a common natural component of the boreal forest.

If jack pine has the only adequate seed supply after a fire on sandy soils, it often becomes the only tree species in the next stand. Similarly, on finer, moister soils, fire will sometimes create pure stands of black spruce. More often, though, what emerge are mixtures of jack pine and black spruce. If for some reason there is a low seed supply of both pine and spruce, however, trembling aspen (called poplar in NWO), which usually develops from root suckers, will sometimes form a pure stand or mix with pine and spruce. If there is a low supply of pine and spruce seed and no live roots of aspen in the burned stand, then white birch can form pure stands or be mixed with any one, two, or all three of the other species. Within any large burned area that covers a variety of soil and terrain, most, or all four, of these species can be found in single and mixed-species stands.

On the former Spruce River Forest Management Agreement (FMA — now part of the Black Spruce Forest), which I managed north of Thunder Bay, Ontario, a total of 11 percent of the forest area consisted of naturally occurring single-species stands of trees and 6 percent was covered by naturally occurring black spruce "monoculture."

Yes, nature produces its own monocultures. On the other hand, it is extremely difficult and prohibitively expensive to create plantation monocultures in the boreal forest of NWO. That is because planting is usually prescribed for the richer soils of timber-harvested areas. On more fertile moist soils, there are often suckering aspen roots and banks of seed from other species in standing trees, logging debris (slash), and humus. On those soils, timber harvesting and treating logging debris and ground to prepare for planting stimulates the regeneration of other species, which then join with those being planted. Later, stand tending seldom aims to eliminate those species and almost never does.

One particular tree stand on the Wolf River Watershed north of Thunder Bay was created after the most intensive and expensive chemical and mechanical vegetation control and ground preparation we ever used. That stand is probably the closest to a human-made "plantation monoculture" that we created in the Spruce River Forest. Today it is approximately 80 percent black spruce and 20 percent a combination of white spruce, trembling aspen, and white birch, with scattered white spruce and jack pine residuals.[1] Nature can, and does, do a more complete job of creating single-species stands, not just after fire, but also after spruce budworm epidemics, which sometimes stimulate regeneration of pure balsam fir stands.

That particular plantation was created during the early years of the Spruce River FMA, when we were trying to earn the public's confidence that we could and would renew and manage the forest in its interest. With experience, we learned that such severe treatment was unnecessary, even on the most difficult sites. By the time of the first five-year audit, we were using lighter machinery and different machines for different soils. We had more confidence in our knowledge of the trees' needs and our ability to satisfy those needs. Costs were dramatically reduced and our treatments continued to be consistently successful and increasingly satisfactory in the views of the public, us, our managers, executives, customers, and investors.

Almost the entire three hundred square kilometres of plantations that were created on my watch contain natural regeneration of other species. Except for more noticeable tree spacing in some portions of plantation-created stands, they are hard to distinguish from most of the natural stands with variations of mixed species. In most NWO forest plantations, the presence of rocks, stumps, large woody debris, and soil variation prevents regular tree spacing. Most of the single-species stands in our forest management legacy are composed of jack pine that we encouraged by spreading logging slash and locally collected seed on clear-cut, sandy soils. The resulting stands contain local genetic stock and resemble the natural, pure pine stands that were harvested and that had regenerated following forest fire.

## JACK PINE — NOT TOTALLY DEPENDENT ON FIRE TO REGENERATE

Because fire is so frequent in the boreal forest, jack pine (*Pinus banksiana*) and some other species have evolved to regenerate and grow following its occurrence. I have seen jack pines as young as five years old bearing seed cones, and once a tree starts producing cones it tends to produce a crop each year. Its cones may persist on the tree and hold viable seed for up to twenty years and in some cases even longer. The volume of seed on each tree and on a given area of ground compounds as trees mature but then declines after the trees reach a certain age. Some cones on some trees open and shed seed on hot, dry days when the resin that seals the cones softens. However, the cones on most live, standing trees remain unopened until the intense heat of a fire melts the resin, thereby releasing a shower of seed.

For several days after a fire, seeds continue to fall, as the still-standing burned trees sway in the wind, lift their roots, and crack any remaining organic soil cover. Many seeds fall into those cracks or find reliable moisture on bare mineral soil, near stumps or other large woody material, or under the edges of rocks. Within days, the seeds germinate and begin to grow. The emerging seedlings' coarse roots aggressively seek out moisture and nutrients. On deeper, dryer soil, they may send roots down to reach more reliable moisture and even the wide-spreading lateral roots may put down additional but small tap roots (sinker roots), thus making the trees relatively wind-firm.

The species requires full sunshine in which to grow. One will almost never notice jack pine seedlings beneath a fully stocked, mature stand of its own or any other species. Even in the partial shade near the edge of an undisturbed stand, only scattered seedlings can be found and most of those perform poorly. This situation is even more pronounced on the shadier north side of a stand. Toward the middle and away from shade at the edge of a cutover or burn, increasing numbers can be found and there they exhibit more vigorous growth.

Because of the species' intolerance of shade, partial-stand harvesting does not encourage jack pine regeneration or seedling growth. However, on clear-cuts natural regeneration can be encouraged by distributing and crushing cone-bearing treetops and roughing up the organic surface soil. Harvesting machinery often does this as it skids logs to roadside. But the ability of such machinery to do so is unreliable, so natural regeneration is usually augmented by systematic, mechanical roughing of the soil (scarifying) and simultaneously injecting seed.

International equipment manufacturers have developed a variety of machines designed specifically for scarifying forest soils and injecting tree seeds. Some patch scarifiers scrape patches of the organic soil surface and mix it with mineral soil. These patches resemble the mix of soils left by the lifted roots of windthrown trees in the natural forest and offer an enhanced growing medium for the developing seedlings.

On Abitibi's private lands north of Thunder Bay, one observant and innovative forest technician/biologist over years of practice studied the trees' response to soil, logging pattern and timing, site treatment, and weather. He worked with a contractor to adapt available machinery to satisfy the trees' needs and to adjust to the ground conditions as together they responded to the correct weather. He consistently delivered thriving young stands at minimal cost that are remarkably similar to natural, fire-origin jack pine stands.

On coarse, sandy soils, and shallow soils over bedrock, jack pine stands generally have more stems per unit of area and trees are smaller, but on fine, rich, moist, deep soils, the species usually exists in mixtures with other species. There, if its neighbours are of the same height or shorter than the jack pine, some trees can become large and make excellent sawlogs and good-quality utility poles.

As seedlings grow into saplings two to six metres tall, their limbs begin to meet and suppress slower-growing jack pines in the stand until the latter die. The resulting stands following fire and clear-cutting are made up of trees all of the same age and with nearly uniform height. Trees that are better suited for pulpwood and sawlogs generally grow between two to three metres from their closest neighbours.

## BLACK SPRUCE — THE BOREAL FOREST'S COMMONEST TREE

Black spruce (*Picea mariana*), like jack pine, evolved to regenerate and grow following fire. It is the commonest tree of Canada's boreal forest and according to the Ontario Ministry of Natural Resources and Forestry (MNRF) "Ontario's most common tree is the black spruce (37.2% of all growing stock)."[2] In NWO, the tree's predominance is mostly the result of frequent large fires, but it is also a result of its adaptability to almost any boreal soil, and its relative resistance to frost and most fungal and insect pests.

"Buds of black spruce open a week or two later than those of nearby white spruce, thus sustaining less damage from late spring frosts."[3] Balsam fir also tends to begin its spring shoot elongation and new needle growth approximately one to two weeks earlier than does black spruce. The developing shoots of black spruce thus tend to also suffer less from severe defoliation by spruce budworms, which emerge with earlier developing shoots of balsam fir and white spruce. At that stage, the tiny budworms' mandibles are too tender to handle the tough, mature foliage of black spruce trees.

After it reaches about ten years of age, black spruce grows seed cones every year, some years more than others, and its cones may hold viable seed for twenty years and longer. Some cones open with low humidity and can shed seeds throughout the year. Others remain closed until the high heat and low humidity from a fire or direct summer sun on the floor of a harvested area opens them and they shed their seed.

I have observed red squirrels systematically nipping off branch tips and whole branches that are heavily loaded with cones on a black spruce tree. They tend to concentrate their effort approximately fifty centimetres below the treetop, where they get the best return for their work. This leaves a tufted and sometimes lopsided top on the tree, and over several decades of pruning by squirrels, black spruce trees develop their characteristic irregular crowns. Those irregular crowns, profiled against a setting sun, are a classic boreal scene.

The species' seeds are much smaller than those of jack pine and in strong wind they travel farther from the parent tree, thereby effectively seeding larger areas. The emerging seedling's roots are finer and less aggressive than those of jack pine. They have difficulty reaching reliable moisture to get started and survive unless they are in some shade, on thin organic matter, or exposed mineral soil. Fire usually provides all three and timber harvesting can be managed to do the same.

During modern timber harvesting, machinery compacts surface organic matter and mosses, thus shortening the distance that spruce roots have to penetrate to access mineral soil. Skidding (dragging) of trees over the forest floor also breaks and thins much of the organic mat and exposes some mineral soil. Subsequent ground preparation (scarification) for seeding of jack pine or planting also exposes more mineral soil. An additional benefit of the machinery is that it crushes and kills existing seedlings (advanced growth) of balsam fir. (There will be more in subsequent chapters about why this is a benefit.)

Roots of even mature trees seldom go deeper than a few centimetres into the mineral soil, and thus provide little resistance to strong winds. Within stands, the roots of several trees are densely woven together and on upland soils, particularly on coarse soils and bedrock, in extreme winds large patches of black spruce may blow over together like dominos.

The species is able to occupy all soils and do well on most of them, except medium to coarse sand. However, it can grow well even on coarse sand if there is a thin layer of finer material on top to hold moisture. Among all of the boreal tree species, black spruce is the best adapted for growing on wet, organic (peaty) soils in muskeg bogs and the edges of fens. Its shallow roots sit at the top of the water table and colonize the growing moss mat. As the mat thickens, the tree grows new roots (adventitious roots) from its trunk near the surface of the moss. This adaptation enables it to reach nutrients found in the living moss, from which it satisfies most of its nitrogen needs, and also evade the rising water table. An excavated root system in such locations may resemble a tap root structure; however, it developed in reverse order to tap roots. Roots below the water table are useful only for increased stability as new feeder-roots grow further and further up the stem in tandem with the growing moss. I have observed such roots in highway rights-of-way cuts, and after river and lakeshore erosion.

In less dense stands, lower branches that come into contact with the moss may also grow roots (layering) and on wet soils many well-formed trees in a mature stand began as layering.

During the early days of the Spruce River Forest FMA there was concern that the paper pot containers in which we were planting black spruce were restricting their root development and predisposing the trees to future instability. We were aware of black spruce's adventitious rooting ability and began planting the seedlings deeper. There was widespread concern with this practice too at the time. We were cautioned that planting seedlings too deeply restricts tree growth. The result was the seedlings grew roots above the pots and today, some thirty years later, the plantations are superior in growth to and apparently as stable as are natural black spruce stands.[4]

One particular natural, former-black-spruce stand, which I have been observing annually since 1985 approximately one kilometre south of the Dorion cut-off on Highway 527 in NWO, has been going through gradual transition. The first time I saw the then one-hundred-year-old stand, it was pure black spruce. The stand extended from deep peat at the bottom of a west-facing, 10 percent slope and continued up the slope, on good mineral soil, all the way to the crest of the hill. Within two years, the trees at the top of the hill had blown over and with passing years the windthrown area extended farther and farther down the slope. In

2016, with a few exceptions, the only remaining standing spruces are on the deep peat at the bottom of the slope. As one travels up slope, climbing over downed and rotting spruces, one moves into more and larger balsam firs which replaced the spruces as they fell.

The stand is an excellent demonstration of how black spruce stands break up and which species replace the spruce after fragmentation. Because of higher amounts of wind exposure at the tops of slopes, trees there tend to fall over before those at the bottom of slopes. Portions of stands on slopes facing the prevailing wind also tend to fall over sooner than those at the bottom. Those at the bottom are less exposed to wind and tend to survive longer. A report on a study in Ontario's northern clay belt states that, "Black spruce stands can be expected to become susceptible to windthrow once dominant stand height reaches 20–21 m."[5] I suspect that black spruces in the northwest also tend to topple once they reach a certain height. It is a reasonable assumption that trees on hilltops and windward slopes may topple at even shorter heights and younger ages.

Unlike its jack pine friend, black spruce can germinate, survive, and grow in the partial shade of most standing trees. Because of this trait, it can, with extreme care to protect it from wind, be managed with a partial-stand harvest system. The tree's shallow roots, however, make partial-stand harvesting a risky venture particularly now with the tendency toward climate-change-induced, stronger, more frequent wind storms. The tree's best protection against wind is the shield provided by its immediate tree neighbours. However, as just described, large sections of a stand in exposed locations will frequently fall together in extreme winds and partial-stand harvesting, by increasing exposure, increases that probability.

Because of its shallow roots, the species doesn't fully exploit the higher fertility of fine, deep, moist soils as efficiently as can deeper-rooted trees. On the other hand, it can efficiently exploit a thin cap of fine-textured soil that catches and retains moisture over rock and coarse sand. Indeed, on lower slopes black spruce stands can do quite well on almost no soil if there is nutrient-rich water seeping down from the higher slope. This can sometimes be observed on boulder trains left on one-time glacial creek beds.

## BLACK SPRUCE — PROMOTED BY FIRE

Fire's greatest gifts to emerging black spruce seedlings are the preparation of a seedbed and the temporary removal of all competing plant growth. Fire gives the seedlings a chance to get started and grow to a size that can withstand most grass and herbaceous competition even on moist, rich soils. Although fire will sometimes establish pure black spruce stands on rich mineral soil, more often it creates a mixture of black spruce and jack pine, with scattered trembling aspen and white birch.

When in those mixed-species stands, black spruce will usually become overtopped by jack pine, which reaches for the sun and gets ahead of the spruce. The spruce, more tolerant of shade, survives and grows slowly beneath the pine. As the pines mature and age, longer-living spruce trees react to openings vacated when some pines die by accelerating their growth and gradually becoming the dominant species in the stand. The short-lived aspen and birch trees take a back seat, having contributed annual leaf litter that enriched the soil and buffered its acidity. As the spruce trees reach old age and die, the stand transitions to more shade-tolerant species, such as balsam fir and white spruce (more about this later in discussion of those two species).

When managing for black spruce, foresters create the conditions that the species demands, by paying careful attention to the variety of terrain, soil, and current and anticipated plant species. In general, the poorer the soil the easier it is to regenerate black spruce naturally; and other-plant competition is less of a threat. That is a lucky coincidence because tree growth rates on poorer soils are generally slower and thus crop yield and return on investment are lower.

Black spruce swamps in NWO will, with due care during harvesting, naturally regenerate well to black spruce. Indeed, a severely machine-rutted fen that I have observed for thirty years, near the Dorion cut-off north of Thunder Bay, today bears an apparently healthy, naturally regenerated black spruce and eastern larch stand. The ruts have partially grown over with sphagnum moss and tree roots but some holes remain. Deep water-filled holes are a natural feature of black spruce fens. Ironically the spruces in this stand appear to be experiencing superior growth to that which was experienced by trees in remnant portions of the original natural stand. A recent visit to that stand further encouraged me. Approximately every ten metres in any direction I was crossing a well-worn snowshoe hare path and there was marten scat on a spruce stump. Obviously prey and predators are finding the new stand to their liking.

The demand for full sunshine in getting black spruce regeneration started is less than it is for jack pine. Shade helps provide relatively consistent moisture and thereby assists seed germination, seedling survival, and early growth. Once the emerging seedling is well rooted, continuing shade depresses growth and from then on the young black spruce tree grows best in full sunlight.

Protecting suppressed seedlings and small trees that were present in the harvested stand can lower renewal cost, and depending on the number and spacing of stems, even reduce the time between harvests and enhance yield. Ensuring availability and distribution of a seed supply after harvest and compaction of the moss and surface-litter layers can enhance seed germination and seedling survival. This prescription can be applied successfully on most waterlogged peat soils and shallow mineral soils over bedrock. As a further bonus these areas seldom require control of competing vegetation, and in my observations white birch mixed with the spruce on shallow upland soils has enhanced the growth of the spruce trees.

On the area that I managed north of Thunder Bay, our black spruce planting was generally confined to richer and moister upland soils where we found planting was necessary to ensure that the species maintained its status in the forest.

In the absence of fire on richer upland soils, with the exception of jack pine and white birch, other trees, shrubs and even herbs and grasses quickly suppress black spruce. This competition prevents black spruce from becoming a major part of the new stand. Careful control of competition on those rich soils is usually required following timber harvesting to get a successful black spruce stand up and running.

Yes, black spruce is the tree that most of us visualize when we imagine the boreal forest. I often catch myself daydreaming of sitting in a canoe gazing at a colourful sunset above a skyline of black spruce silhouettes. I find the scene relaxing as I conjure abstractions of infinite different black spruce forms from my memories of time spent on eastern Canadian boreal lakes. Today I communicate those images of my favourite tree in ink drawings on the bark of my next-favourite tree, white birch.

## WHITE BIRCH — TWO IN ONE

There are two species of white birch trees in NWO, mountain or eastern paper birch (*Betula cordifolia*) and paper or canoe birch (*Betula papyrifera*). Eastern paper birch ranges from just east of the Ontario-Manitoba border east to Newfoundland and south into the northeastern United States. Canoe birch covers the same range but also extends west to British Columbia and north to Alaska with the exception of the Pacific coast.[6] In NWO, both species are often found growing adjacent to each other.

Eastern paper birch has relatively thin bark that tends to be constantly shedding in strips and can be easily separated into layers. Tree trunks usually have numerous strips hanging in tight to loose curls, giving them a ragged, unkempt appearance.[7] The bark's exterior colour varies in horizontal bands between white, grey, pale yellow, and even a very pale blue, and beneath recently shed strips it can have a salmon shade. The inside-bark surfaces are usually bronze to copper coloured but I have also seen shades of tangerine and burgundy. It can more often be found on higher slopes within any given location giving way to canoe birch on lower slopes and in valley bottoms.

Canoe birch, as you suspect, is the tree from which birch bark canoes are made and its main defining characteristic is its relatively thick tough bark that, during a short period in late spring to early summer, can be peeled in large, intact, waterproof sheets. Its bark tends to be milky white on the exterior surface and soft pink on the inside surfaces of difficult to separate layers.

Most recognized authorities report that the time of seed dispersal varies from July to late fall. Late in March one year in Newfoundland, my crew and I were broadcasting spruce seed over the snow on a previous year's burn. The area had been clear-cut of coniferous trees before the fire but still had scattered, large, living birches. While checking small sampling plots to determine if we were achieving the desired distribution of seed, I noticed that birch seed and bracts had accumulated at the bases of snow drifts. That suggested to me that they had fallen during, or after, the most recent snowfall.

Both birch species can be found in pure stands or associating with any of the other boreal tree species. They are intolerant of shade and are often among the first trees to get started after any stand-removing disturbance. A severe fire, spring, summer, or fall harvesting with soil disturbance will promote regeneration from seed. By far the best seedbeds are bare mineral soil, mixed mineral and organic soil, and rotting wood.

One will often see several tiny white birch seedlings on the moss-covered top of an old stump or log in a small opening of a stand. These seedlings seldom develop into trees unless the opening enlarges, giving them access to full light.

One particular birch tree stands out in my memory. Apparently the seedling started on the top of a rather high old stump that had long since rotted away. The last time I saw the tree was in 2006 when the healthy maturing specimen was standing atop several finger-like roots that still outlined the now-missing stump on which it had originated.

A light fire or winter harvesting in young stands will promote sprouting from the bases of stumps. Sprouting is less vigorous as trees age and in my observations has been rare from trees that were older than sixty to seventy years.

White birch is relatively short lived but some stands on deep, moist, rich soils, which avoid fire for up to 120 years, will still contain a few relatively large, healthy trees. They experience severe shock after significant change in their immediate environment. Depending on a birch tree's location, the removal of even one of its neighbours can sometimes precipitate crown decline and eventual tree death.

Birches suffered severe dieback, causing tree mortality, all across eastern Canada between 1930 and 1950. Since then they have recovered well. Unfortunately there is currently similar wide-scale dieback and tree mortality in Minnesota and I have seen pockets of mortality as far north as Thunder Bay. I am not aware of any clear consensus among forest scientists on what the causes of this dieback are but some point to repeated insect attacks and drought weakening the trees and making them susceptible to other insects and root rot. One pathologist I knew several years ago isolated a fungus that was common to declining birches but I think he was uncertain whether it was the cause or effect of decline.

For millennia, First Nations people have made use of birch trees possibly more than they have any other tree species. They used birch for purposes as varied as home construction, making water craft, snowshoes, cooking pots and utensils, and a variety of artistic items. The bark, roots, and smaller stems were most useful. In NWO and Minnesota making birch bark baskets is a revived craft and during 2014 both Muriel and I took a course in it at the North House Folk School in Grand Marais, Minnesota.[8]

Since 1980 I have been drawing with black-and-white inks on the outside surface of birch bark. I see partial forest scenes in the medium's colour and texture variations and enhance them to illustrate niches in the boreal forest. The framed results have found their way into homes and offices around the world. At $0.37 to $0.75 per square centimetre ($2.50 to $5 per square inch), I figure that I am adding decent value to an otherwise undervalued raw material.

Large, clear-wooded white birch logs from the best growing sites are normally in demand for veneer because of the wood's smooth texture, colour, and grain, which appeal to home-owners and artisans. In recent years birch has been used in hardwood kraft pulp, specialty saw products, and oriented strand-board. Today pulp and paper mills are increasingly harvesting low-quality birch from which they generate some of their energy needs. The later practice is forcing some northern residents who heat their homes with birch firewood farther afield for their supply, causing growing hardship and resentment. It behoves forest managers to take note and work with firewood users to ensure an economic supply and improve community relations.[9]

## TREMBLING ASPEN — COULD IT BE EARTH'S OLDEST LIVING THING?

I believe that we all enjoy the sound of wind moving through tree crowns. Each species tends to have its own distinctive sound because of variations in branch and leaf texture. Even a gentle breeze passing through the crown of a trembling aspen tree (*Populus tremuloides*) disturbs its brittle leaves on their long, rather-flat stems; as they tremble they bump against each other, creating a clicking, rushing sound, hence the tree's common name. That sound is a characteristic of the boreal forest from Alaska to Newfoundland, but it can also be heard throughout the species' range well south into a variety of other forests in the United States. I have even seen it in the desert mountains of Arizona at elevations known as "The Canadian Zone." It has been leafless during my winter hikes and climbs so I could not enjoy that familiar sound of home.

Large trembling aspens with their coarse upper branches are favourite nesting trees for crows, ravens, and a variety of hawks and owls, which build large nests made with sticks. Pileated woodpeckers and other boreal woodpeckers frequently build their cavity nests in aspens and other cavity nesters from chickadees to wood and golden-eye ducks use the woodpecker's abandoned nests.

One of my favourite sounds is that of a woodpecker's hammering as it echoes through an otherwise silent forest. The loud staccato of a pileated woodpecker chipping chunks of wood from the base of a trembling aspen or any other large tree is a common sound bite of the Ontario boreal forest. They are usually after carpenter ants that create numerous tunnels, which eventually weaken the tree until it falls over.

While sitting in my canoe fishing for speckled trout one calm summer evening I was startled by a loud crash on shore. I was certain that a tree had fallen. Intrigued I decided to investigate. Less than thirty metres up a steep, fertile slope, a trembling aspen half a metre in diameter had moments before falling over. There were only five centimetres of sound wood remaining around the tree's stump perimeter; inside, the wood looked like Swiss cheese and was teeming with carpenter ants. Distinctive, loonie-sized chips beneath the butt of the fallen tree were proof that a pileated woodpecker had earlier been doing its duty. Unfortunately for the hapless tree it was too late. The woodpecker's "felling notch" and the continuing feeding of the ants were more than the weakened stump could bear and it had collapsed beneath the tree's weight.

One of my favourite spring sights is a female bear feeding with her cubs on the seed catkins of a mature aspen tree. I remember one spring accompanying a publicly vocal critic of my forestry practices to see things as they were. As we were driving through one of our largest clear-cuts she exclaimed, "What in the heavens is that?" There in the top of a large reserved (i.e., a tree left uncut by timber regulations) aspen tree in the middle of the clear-cut was a female bear and her three cubs pulling branch tips toward them, stripping catkins with their long tongues. During early spring aspen catkins provide nutritious food when it is otherwise in short supply for bears. The conversation immediately took a more positive turn.

If you have ever seen a collection of broken branch tips in the top of a mature aspen it was probably created by a feeding black bear. Such collections are referred to by some bush-wise persons as bear nests.

Among the boreal trees, trembling aspen is unique in that its main means of regeneration is via root suckering instead of seed. The suckers develop from buds on an extensive root system of apparently different trees. In fact the trees are all one organism of the same genetic material, or clone. "Groups of clones derived from a single aspen seedling by repeated generations of root sprouts have been found that occupy up to 80 hectares and consist of thousands of trees. These clones may have originated on land exposed soon after the Pleistocene ice sheet melted, making them among the largest and oldest organisms in the world."[10]

Different clones have different growth characteristics. The leaves of a single clone will emerge in spring and change colour in the fall at the same time but possibly at a different time from those of neighbouring clones. Next spring notice how on one hillside aspen leaves have emerged but on another adjacent hillside the aspen trees will still be bare of leaves.

Aspen, generally called poplar in Ontario, demands near-full sunlight to get started and must remain close to the height of its immediate neighbours to survive and grow. It therefore thrives in even-aged stands. However, if an even-aged stand of aspen has for years been severely defoliated by tent caterpillars, and some trees have died, aspen will on some soils regenerate from suckers in the openings and beneath the thin crowns. The stand will then briefly exhibit a somewhat uneven-aged structure until the remaining original trees die and the new stand becomes even-aged.

Trembling aspen is another pioneer species that on most upland mineral soils, when given the chance, will aggressively replace the previous stand of any boreal species following a variety of disturbances. However, severe fire tends to reduce that chance, unless the seed sources of other species were absent before or were consumed by the fire. Aspen seeds can travel long distances enclosed in their light-weight cotton balls, even on light winds. Over several years, they, like white birch, can regenerate where other species have failed. During the first ten years or so after planting of any other species, aspen in-growth from off-site seed can occur and in short time severely suppress growth of the planted trees.

Following clear-cut timber harvesting, immediate oc-cupation of the cutover by a dense aspen-sucker stand of-ten causes nightmares for foresters, particularly if their government-approved forest-management plans scheduled regeneration of another species. Without intervention to dra-matically reduce the number of aspen stems, its dense canopy of leaves will restrict development of other light-loving species and favour balsam fir and white spruce as part of the next crop. Even black spruce seedlings have difficulty surviving beneath a dense young aspen thicket.

In NWO, aspen stands seldom burn except by ground fires before spring green-up. If one does burn hot enough to kill the standing tree, it will usually quickly regenerate to another aspen stand because such fires are seldom hot enough to damage the aspen roots.

The tree's normally unblemished white wood, resistance to splintering, and tall stem with low taper has gained popularity as lumber. It is now also a raw material in manufacturing pulp, window, door, and wall trim, and a variety of reconstituted sheeting and flooring products. Some sauna owners boast of their splinter-free bottoms for which they give credit to their aspen sauna benches.

During my career aspen has moved from the status of weed to that of a desired species because of increasing demand for wood and changing technology. We are still going through one of the longest market slumps in history for wood, but I expect that the demand for all wood will continue to increase over the long term as we develop yet unimagined uses for this versatile and renewable raw material.

## BALSAM FIR — GOOD AT FILLING FOREST VOIDS

Balsam fir (*Abies balsamea*) is my favourite Christmas tree. I love the aroma that fills my house as soon as it enters the door and stays until it is removed after my family's celebrations end on the traditional twelfth day of Christmas. Balsam fir is, however, seldom a forest manager's favourite tree.

We have all heard the saying, "nature abhors a vacuum." Balsam fir has evolved to fill vacuums that other trees may not be adapted to fill under prevailing conditions. Those spaces could otherwise be void of a tree if firs were not around, and it is almost always around in the absence of fire, ready to seize any opportunity. The tree will occupy any soil, does well on moist but not wet sites, and performs poorly on dry soils.

Balsam fir's range extends from Alberta to the island of Newfoundland but, because of its preference for moisture, it is commoner and more vigorous east of Ontario. It made up 47 percent of the forest that I managed in Newfoundland but only 6 percent of the forest that I later managed in Ontario.

Seed cones are produced annually on open-grown or dominant trees after they reach about fifteen years of age. The cones ripen and fall apart in September and shed all of their seed. Usually this prevents balsam fir from regenerating following fire as any fallen seeds from previous years and established seedlings are burned. However, occasionally a dense stand consisting almost entirely of fir can result from an early September fire after seeds have ripened and before seedfall. While working in Newfoundland I examined a documented stand that originated after a September 1921 prescribed fire in balsam fir south of the community of Badger and southeast of Pamehoc Lake. In 1968 the resulting near-mature stand was almost entirely balsam fir. In 1985 I observed a similar stand south of Garden Lake on the Spruce River FMA north of Thunder Bay. My search located a couple of badly decayed but charred stumps, indicating that the previous stand had burned. I concluded that the stand could only have originated following an early September fire in another balsam fir stand.

Fir's relatively large seeds can germinate and its coarse-rooted seedlings can survive on moss and forest floor litter in deep shade. Close examination of the forest floor under most stands in NWO will reveal some tiny balsam fir seedlings, sometimes tens of thousands per hectare. They are usually less than ten-centimetres tall, but may be up to twenty years old. As the over-topping stand of any species matures and individual trees die, they create openings in the stand canopy. The little fir seedlings then quickly begin growing and fill the openings, thus making it difficult for light-demanding species to regenerate and grow.

Under some black spruce stands dense feather moss on top of thick organic matter restricts fir's regeneration and survival, but even in these cases as trees blow over and their roots lift and crack the organic mat, fir eventually infiltrates.

I emphasize that balsam fir is very shade tolerant and possibly the most shade tolerant of all of Ontario's coniferous trees. In the continuing absence of fire most jack pine and black spruce stands gradually convert to balsam fir mixed with other shade-tolerant species such as white spruce and some semi-tolerant black spruce. After spring or summer fires in balsam fir stands in NWO, the new stand will regenerate to something else. If only a few jack pine and black spruce per hectare are present, the new stand can be predominately composed of those two species, with some trembling aspen and white birch. If seed-bearing pine and spruce are absent in the burned stand, the new stand may regenerate to white birch or trembling aspen, or a mixture of both, or in extreme cases, depending on soil, to hardwood shrubs.

Now let's look at how tiny fir seedlings beneath maturing stands can cause problems in stand management and have far-reaching consequences on future plant and possibly even wildlife composition. They even eventually impact the economics of wood delivery and mill efficiency.

## BALSAM FIR — TOO MANY OR TOO FEW?

Normally one thousand or more well-distributed coniferous trees will fully utilize a hectare of ground in the boreal forest. Now we look at how thousands of little fir seedlings per hectare can create a dilemma for forest managers who are trying to sustain the prevailing natural stand pattern and species distribution.

Early timber harvesting across eastern Canada using manual felling and extraction with horses usually resulted in the survival of most tiny fir seedlings. The relatively gentle disturbance to only a small portion of the forest floor ensured that the majority of fir seedlings remained undamaged. Additionally the limited disturbance of the organic mat did little to create seedbeds for black spruce and jack pine. Consequently the "horse-logged" areas, especially on medium to rich soils, converted predominantly to fir. I maintain that the fir stands that originated because of horse logging enhanced the severity and extent of the large spruce budworm epidemic of the 1970s and 1980s in northern Ontario.

Modern machinery used in clear-cut harvesting crushes most of the advanced-growth fir seedlings, and this dramatically reduces their numbers in the new stand. The same machinery enhances new spruce regeneration by compacting mosses, plant litter, and organic surface soils helping any emerging black spruce germinants' weak roots to reach moisture in underlying mineral soil.

In partial-stand harvesting, where machinery is confined to narrow trails, firs will usually replace the harvested trees much as they did in horse-logged areas and will probably eventually dominate the stand.

During the 1990s foresters at Abitibi-Price did a roughly sixty-hectare partial-stand harvest (selection harvest) in a mixed jack pine and black spruce stand on the company's private lands south of Dog River, Ontario. The selected trees were felled and processed into pulpwood and sawlogs using a modern cut-to-length, single-grip harvester, and forwarded to roadside along marked trails by the system's tracked forwarder.

I observed that, within five years, many of the reserved trees had been windthrown, and that the emerging new stand was predominantly balsam fir. I had seen that result from various commercial thinning trials my employer, Price (Nfld.) Pulp & Paper Limited, and its predecessor Anglo (Nfld.) Development Company Ltd. (AND), had carried out in Newfoundland from 1921 to 1968 and was not surprised to see the same in Ontario. That partially explains why I have in the past and, if practising today, would still prescribe clear-cut harvesting in the boreal forest of NWO.

As balsam fir grows it becomes a favourite food for a variety of insects such as the spruce budworm and, in the more eastern provinces, particularly on the island of Newfoundland, also hemlock looper and balsam woolly aphid. In most areas it is difficult to maintain a balsam fir stand into old age as one or another of these insects will usually thwart the best laid plans of any forest manager. Even in the absence of insect epidemics, several species of fungi cause increasing wood rot, reducing quality and yield. Where fir is a large component of a stand, eastern forest managers tend to harvest it between forty and fifty years old, before the insects and fungi take their toll. If NWO weather becomes warmer and wetter and fir increases in the forest it is possible that all three of the mentioned insects will cause problems.

Some Ontario industrial forest managers consider balsam fir a weed because of the high risk one has to accept when managing it as a crop. It occupies valuable growing space that can be used by more desirable and manageable trees and, to some managers, it becomes a nuisance when it arrives at their mills.

There is good reason for this perspective. Fir's wood is weaker and more brittle than that of black spruce, particularly when it is dry, and it has significantly lower wood density than do jack pine and black spruce. These factors cause problems on paper machines, lower pulp yield from the wood, lower mill efficiency in a number of ways, and higher cost in both wood delivery and processing in the mill.

Where balsam fir is a major part of a pulp mill's wood supply, owners can compensate for those apparently negative characteristics with high strategic investment. During the 1960s, Price in Quebec City developed a wood-chip refining system that was an improvement over conventional stone grinding of logs for making mechanical pulp. The new technology was installed along with a new state-of-the-art newsprint machine in their mill in Grand Falls, Newfoundland.

That strategic investment converted the mill from one that had used roughly 30 percent chemical (sulphite) pulp and 70 percent stone-ground pulp into a user of 100 percent mechanical refiner pulp. In doing so the mill reduced its effluent, increased yield from its wood, improved the printability of its newsprint, and, to the best of my knowledge, reduced the negative impact of fir's low wood density. The conversion was economically possible because the company generated a major portion of the high electrical power requirements of the new process. A former manager of the mill tells me that eventually the mill had to inject some chemical into the process to meet some customers' specifications. Because of superior paper strength and printability, higher yield from wood and less effluent, the conversion was a success, but not entirely up to expectations. The mill, despite producing superior paper, was a casualty of the declining demand for newsprint and, at the time this book was being written, was being demolished.

In NWO the relative low and scattered presence of fir in the available timber supply and the high risk involved in its management, in my opinion, does not warrant adapting a pulp mill to utilize more balsam fir, especially with our high energy costs.

Nor is balsam fir often a sawmill manager's preferred wood. There the logs and sawn lumber require segregation from the other species right up to the drying kiln, where a drying regime that is specific to fir is required. In NWO the small proportion of fir in the log supply requires expensive handling and machine down time for what some see as an inferior product with a lower return on investment.

## WHITE SPRUCE — THE OTHER SPRUCE TREE

White spruce (*Picea glauca*) can be found throughout the Canadian and Alaskan boreal forest. "Often found at the Arctic tree line,"[11] this shade-tolerant tree is able to regenerate beneath the canopy of almost any other tree species where it remains suppressed (growing slowly) and responding to increased light when canopy openings occur. Dense understory stands of hardwood shrubs, such as mountain maple (*Acer spicatum*) and beaked hazel (*Corylus cornuta*), beneath aspen overstories on rich soils do, however, prevent white spruce regeneration. I don't recall ever seeing white spruce, and only rarely balsam fir seedlings, under such cover.

White spruce is able to survive several periods of suppressed growth as competition from other trees increases, and can respond with increased growth each time a neighbour dies. This characteristic makes it more manageable in partial cut or selection-harvested, multi-aged stands.

Seed cones ripen in August and almost all seed is dispersed from September through fall and winter. Some empty cones may remain attached until the following spring.

I am often asked how one can distinguish white from black spruce. Generally black spruce bark is greyish brown to brown and covered with small brown scales. White spruce, particularly at younger ages, has a greyish bark that is smooth and may have no scales, but as the tree ages its bark also tends to become darker, browner, and scaly.

Seed cones of black spruce are hard and inflexible at all times whereas those of white spruce are softer and flexible. Black spruce cones are purple before ripening and then turn brownish, whereas white spruce cones are green before ripening and after ripening also turn brown.

Twigs of black spruce have fine orange to brown hairs at the base of needles (leaves) whereas those of white spruce are usually hairless. Seedlings of both species can have hairy twigs. White spruce seedlings at the end of the growing season are stiffer and terminal leaders are thicker than those of black spruce.

The stiffness of white spruce's seedlings and its coarser, more aggressive roots give it a relative advantage over black spruce in some harvested sites where grass has gained dominance. Snow collapses tall grass stems into dense mats that bend, bury, and crush pliable black spruce seedlings; this effectively prevents them from succeeding without human intervention. Stiffer white spruce can regenerate, resist bending, breaking, and becoming buried, and succeed where most other conifers have a tough start. It is this characteristic that has given white spruce the name of "Old Field Spruce" in the Maritime provinces. Blue joint grass (*Calamagrostis canadensis*), because of its high stand density and tall stems, does, however, cause breakage and mortality to planted white spruce seedlings unless relatively large planting stock has been used.

Crushed white spruce needles have a pungent odour that to me is somewhat like that of cat's urine, whereas black spruce needles smell rather spicy and pleasant. Black spruce often excretes an aromatic gum from wounds in its bark. This hardens and as a boy I routinely sucked on a chunk until it softened and then chewed it as a gum. I continued to do so occasionally well into my middle age. White spruce wounds produce a sticky white to yellow resin that smells like its crushed needles and I have never considered chewing that!

White spruce behaves a lot like balsam fir and the two grow together in mixed stands, sometimes with black spruce, white birch, and aspen, which they gradually replace if otherwise undisturbed by whole-stand removal. In the absence of fire, jack pine and black spruce stands on richer moister soils also gradually convert to a mixture of fir, white spruce, and maybe some black spruce.

As stands dominated by fir and white spruce pass approximately forty years old, they tend to experience insect epidemics after which most of the firs die, but some taller and fuller-crowned firs and white spruces may survive. The dead firs are quickly replaced by more firs and a few white spruces. The resulting stand will be predominantly of one age but with some scattered older individuals until the almost inevitable wildfire occurs, the stand is again attacked by insects, or it is harvested. White spruce is severally defoliated during spruce budworm epidemics, which tend to occur every thirty to forty years, and is almost as vulnerable to those epidemics as is balsam fir. Often large, full-crowned white spruce that survive a spruce budworm attack soon succumb to bark beetle attack.

Fir and white spruce start spring growth at the same time and in sync with the tiny budworms' needs whereas black spruce's growth is delayed by up to two weeks. The budworms' tender mandibles are incapable of feeding on old foliage at that stage so they starve trying to feed on black spruces. As the budworms grow and their mandibles toughen they hang on tiny filaments that they produce, much like a spider's web, and wind blows them from one tree to another. In that way black spruce become re-infested and the larvae are now capable of feeding on their foliage. Black spruce in Ontario seldom succumbs to budworm feeding.

Like balsam fir, white spruce will sometimes regenerate following a late-summer fire if seeds have ripened and cones have not already opened. In the discussion about balsam fir I described a fire-origin fir stand that I saw. Adjacent to that stand there was a white spruce stand that had apparently regenerated after the same fire. I speculated that the stand that had burned had probably avoided fire for well in excess of a hundred years as it gradually converted from jack pine and black spruce to balsam fir and white spruce. I further surmised that before the most recent fire spruce budworms had killed all but the largest of the trees, which in turn supplied the seed for the current stands.

We plant few white spruce compared to black spruce in the northwest because of the former's greater vulnerability to insects and late-spring frosts. Plantations in bottom-land frost pockets do poorly for several years, until the tops gradually get above the frost level. When planting white spruce in NWO some foresters select cooler north slopes. There, active growth tends to start later in the spring, and when late-spring frost does occur the heavier cool air slides down slope below the planted trees thus decreasing the chance of damage.

If the winter of 2012 is a harbinger of future weather events, we will be planting even fewer white spruce in the future. That year the species suffered severely from an extended abnormally mild spell in March during which some trees, especially white spruce, broke dormancy and active growth started. The mild period was followed by a return of winter conditions that killed elongating shoots and the previous year's foliage resulting in widespread tree deformity in white spruce plantations. I observed at some locations that, even among white spruce in mature, mixed stands, scattered tree death occurred the following summer.

White spruce is in demand as sawlogs because of its generally larger size than black spruce. However, when given a choice, pulp mills will take black spruce instead of white spruce because the wood density of white spruce is lower, resulting in lower pulp yield and thus higher manufacturing costs.

## EASTERN WHITE CEDAR — PRIZED IN NATURE AND CARPENTRY

Eastern white cedar (*Thuja occidentalis*) ranges from southeastern Manitoba and a small pocket north of Lake Winnipeg to mainland Nova Scotia and Anticosti Island in Quebec. It is also in all the American states bordering Canada between Minnesota and Maine.[12] It is not a prominent species anywhere within the boreal forest. Within Canada it is most significant in New Brunswick and the southern parts of Quebec and Ontario. In the boreal forest it is usually relegated to finer and damper upland soils, along creek beds and in swamps, but occasional trees can be found hiding anywhere mixed with other tree species.

I well remember when I was a student forester and I was timber cruising in both provinces along the New Brunswick and Quebec border. I had to walk for hundreds of metres along a compass line through dense cedar and at times had to lie on the interlocking branches to force my way through. At that time cedar was not considered an important commercial species and we were thankful not to have to measure the trees but rather simply record the distance being passed through cedar thicket. The only location I have seen anything resembling that in NWO is near the community of Dorion.

White cedar is a relatively short tree with a rather stout lower trunk that is often curved, particularly near its base. Another distinguishing feature is its stringy bark; on older trees it forms long, vertical strips that make the bark easy to remove. The trees start producing seed cones when they are as young as six years old and continue throughout life with peak crops every two to five years. The cones ripen in August and seeds are distributed through September. For this reason the tree seldom regenerates well following fire except along burn edges where some unburned cedar may remain, particularly along creeks or swamp edges. Occasionally a summer ground fire will stimulate regeneration if the small developing cones aren't burned. In swamps regeneration is usually via layering — that is, through rooting of low-lying branches that come into contact with the ground or its moss cover. The rooted branch becomes a new tree and it is that beginning that causes the curved lower trunk of many swamp cedars.

White cedars can survive for several hundred years but seldom without severe heart rot and on wet ground many develop hollow trunks. Dried cedar wood, however, becomes very rot resistant. The wood splits easily and is light weight. Thirty years ago, I built a split-rail fence of cedar around my front yard. Three years ago, I started replacing some of the posts but most of the rails are as sound as the day I put them there. I have enjoyed observing and photographing the diverse communities of lichens that have formed on some of the rails. They have also become popular roosts for a variety of seed-eating birds, which have blessed me with a large variety of plants that continue to appear and multiply beneath the rails. To avoid having my house become hidden in a forest I recently had to remove seven different species of trees and shrubs along one ten-metre length of the fence.

Because cedar is an easily split, durable wood and has shredding bark, artisans can manufacture useful items while using a minimum of specialty tools. First Nations peoples used cedar wood as planking in their birch bark canoes and in basket making. The bark was used to make ropes, fire tinder, and a host of other practical items. Early European immigrants later used the easily split wood for roofing, fences, barrels, kindling, and a variety of other items to make life more comfortable. Many birds and mammals cherish the soft, fibrous, inner bark as lining for their nests or burrows, which are often contained within a hollow cedar tree.

Today cedar is still prized because of its easy workability, grain pattern, durability, and aroma in a variety of products and structures. Most well-designed buildings that effectively use cedar have become eye candy for those of us who appreciate natural wood construction. My prize is a three-season "gazebo" that I started in 2008 using local cedar and finally completed during 2011, thanks to persistent, mischievous pestering by a neighbour.

## EASTERN LARCH OR TAMARACK — HAS UNIQUE FIBRE AND CHEMISTRY

Eastern larch is also known as tamarack, hackmatack, American larch, and juniper, but its scientific name is *Larix laricina*. It is found in all treed areas of Canada (except west of the Rockies) and extends north into the Arctic.[13] Around population centres it is often confused with any of three exotics: European (*Larix decidua*), Siberian (*Larix sibirica*) and Japanese (*Larix kaempferi*) larches and their hybrids. All three and their hybrids have been planted as ornamentals in many residential areas, and in a number of forestry trials in northern Ontario. In some locations that I have observed, the exotics appear to be naturalizing and expanding their ranges.

I distinguish eastern larch from the many exotic larches by its needles, which are rarely more than two to three centimetres long; those of the exotics are often over four centimetres long. The seed cones of all three exotic species also have those different lengths. All larches are deciduous and their needles turn a brilliant gold after the leaves of most other deciduous trees have fallen. I sometimes stand transfixed on a frosty late-October morning marvelling at the brilliant gold and red tapestry of larch, blueberry, and other heath plants on recently harvested areas and muskegs.

Often relegated to wet and peaty creek beds and swamps, this shallow-rooted species also does well on a variety of upland soils and can even be found as stunted, twisted specimens anchored in cracks on otherwise bald rock. It is very shade intolerant and is often a partner with black spruce near the edges of burns where it overtops spruces of the same age. Unlike spruce under jack pine, the spruce under larch grows relatively unimpeded by the partial shade of the larch's thin foliage.

Larch shed their seeds from mid-August through the fall and the seeds germinate and survive best on wet sphagnum moss or bare mineral soil. That is why larch are so prominent along highway rights-of-way and the shoulders of many bush roads in the northwest. Like eastern cedar, eastern larch has not been as popular a commercial species as the spruces and jack pine but it does have its fans. An acquaintance of mine in Newfoundland owns a small specialty sawmill complete with a planer and kiln. He manufactures white birch and larch flooring as well as larch wainscoting and trim. When we last talked a few years ago, he had a good market for his product. I toured his house that he had finished with larch and was impressed with his taste because of the rich pattern and colour of the finished woodwork. Many sawmillers complain that larch is difficult to dry without severe warping but he appears to have found the right drying regime as I could see no evidence that he had experienced any warping problems.

Larch was prized for centuries by Newfoundland boat builders who used it in keels and planking because of its high durability and resistance to abrasion; an important quality for boats that spent much of their time battering and being squeezed by Arctic ice floes. A Dryden, Ontario, senior-forester friend of mine reminds me that larch also makes excellent decking because of its rich colour, durability to abrasion, and high resistance to rot, thereby requiring no preservatives. We both recently reminisced about its excellence as firewood but remembered how it burns, so hot that cast-iron grates and stoves didn't last long when we burned dry larch. I once supplied a potter acquaintance of mine with a pickup load of dry larch for his pottery kiln so he could get the higher heat he required for his particular type of pottery.

I expect that as demand for wood increases and science finds new uses for larch's unique chemistry and fibre structure, which to now have been impediments to its commercial use, we will see more larch being utilized. With increasing demand we will manage for larch, but its vulnerability to frequent epidemics of an insect known as the larch sawfly will need to be addressed before large investments are considered.

## EASTERN WHITE PINE — NOT A BOREAL SPECIES BUT PRESENT

Eastern white pine (*Pinus strobus*) is found from the southeastern corner of Manitoba to Newfoundland and well down into the eastern United States.[14] This magnificent tree has limited range within the NWO boreal forest, existing mostly along its southern fringe where the boreal and Great Lakes–St. Lawrence forests merge. There are, however, small stands and scattered trees well north in isolated micro-climates, giving the impression that it is a boreal species.

An examination of the tree's needles will reveal that they are clustered in bundles of five on the twigs. This distinguishes it from the other native pines in our region, jack and red pines, which have only two needles per bundle. Its hanging, ten- to twenty-centimetres-long, flexible seed cones, which require two years to mature, begin appearing at about age twenty years. Seeds are "disbursed by wind and seed-eating animals" upon maturity in late summer.[15]

"Rapid-growing; thrives in full sunlight; seedlings moderately shade-tolerant; can survive under an open canopy and attain full vigor if the shade is removed within 20 years."[16] Indeed, partial shade has been shown to offer the tree some protection from the white pine weevil and white pine blister rust, its two most dangerous pests in NWO.

The larva (worm) of the weevil burrows up through last year's main-stem growth leader. Some authorities suggest that, under shade, that leader is too thin to accommodate the worm's diameter, thus limiting its development.

Blister rust requires additional hosts, the gooseberry and currant genus members (*Ribes*), from which it releases heavy spores that must find a receptive pine needle. Dew on needles is considered an enabler because it helps by initially attaching the spores to the needles. Dew formation is less likely under a partial tree canopy. Also shaded young pines shed the needles from their lower branches earlier than do more open-grown trees and are thus less likely to collect heavy spores on higher needles and thus become infected.

The disease spreads from the needles through the twigs and branches into the main stem, causing large sores that may kill the bark and eventually girdle the tree's main trunk. That portion of the tree above the girdle then dies. An internet search will reveal most of what we know about the disease today.

Beginning in the 1990s Abitibi-Price (after 1997 Abitibi-Consolidated) planted white pine mixed with red pine and black spruce on its privately held Grand Trunk Pacific blocks in Ontario around Athelstane and Muskeg Lakes and the Dog River, until the firm sold the land in 2005. It was similarly introduced to the Spruce River FMA north of Thunder Bay and as far north as the TransCanada pipeline.

Semi-mature and mature white pines were already components of some stands on the company's private lands and in the Wolf River watershed of the Spruce River FMA. On both land tenures the company reserved (i.e., did not cut) the white pines, but harvested other tree species (shelterwood harvesting), and treated (site prepared) the ground to encourage natural regeneration of the pines. On the private blocks, in response to geneticists' advice, additional white pines were planted to enrich the genetic mix as successive generations of white pine deteriorate if there is excessive inbreeding.

Subsequent visits during recent years to the planted and naturally regenerated white pine areas have convinced me that the developing stands are healthy, and I find them a welcome addition to the boreal landscape. Such stands should do well as our climate continues to warm.

White pines that have matured in a forest in which the ground is fully covered with trees (fully stocked) tend to have no lower branches. Subsequent wildfires seldom climb into the top surviving branches, which can be metres above neighbouring boreal species. Older trees also have thick, fire-resistant bark that enables them to survive most fires.

Various authorities report that living trees will spread seed from thirty to two hundred metres of the parents following different disturbances. Although the tree survives and grows well following fire or harvesting, the boreal forest's less than optimal climate for the species stresses the tree, rendering it vulnerable to insects and diseases. These factors limit its size, distribution, and occurrence on the landscape.

One peculiar tree of the species has attracted the attention of two of my acquaintances and me. Yes, Haida Gwaii had its Golden Spruce but we know the location of a Golden White Pine. We are cautiously holding the tree's location secret as we attempt to get it protected by the landowner.

White pine never was a large part of the commercial timber harvest in the Ontario boreal forest and it is even less so today. During the 1980s, in response to rising public concern for the species' continuing survival, it became increasingly difficult to receive a licence to harvest white pine on Crown lands. Maybe someday as a result of past and current forest management and changing climate, Ontario's Provincial Tree will also represent what is now the boreal forest.

## BLACK ASH — A VALUABLE TREE IN DANGER

Black ash (*Fraxinus nigra*) is found from southeastern Manitoba to western Newfoundland, south into Iowa and east into the New England states.[17] In NWO it forms only a small fraction of the overall forest cover and is found mainly in swampy areas, near streams, and on upland depressions, and even there it is seldom very extensive.

It is the only boreal deciduous tree with compound leaves, meaning each leaf is made up of several small leaflets that are arranged opposite each other along the leaf's main stem. Where the leaflets connect to the leaf's main stem there is usually a small bunch of brown hair that helps distinguish black ash from green ash, which has generally hairless leaves. During the leafless seasons black ash's stouter twigs with their dark brown to black leaf buds arranged opposite each other, and its soft corky grey bark on young trees and branches separate it from other boreal deciduous trees. In spring black-looking flower clusters can be seen at the tips of twigs in the tree's crown. Black ash is common along the southern edge of the boreal forest into the Great Lakes–St. Lawrence forest south and west of Thunder Bay, where it mixes with green ash. Green ash extends across the prairies.

Black ash has a straight-grained wood that is easily split but when dry the wood is hard and durable. I am the proud owner of three attractive art easels made from black ash by Jan Luit, a Thunder Bay artisan and wildlife photographer. The wood colour and grain complement the natural colours of my birch-bark pictures when displayed at my shows. I know a couple who are eyeing a particular senior tree on their woodlot and making plans to saw it into finished cuts for as yet unforeseen furniture or cabinets in their home. It pains me when I see this beautiful wood being used as firewood, as is too often the case. The species is widely used by First Nations people, particularly for making snowshoe frames, toboggans, canoe paddles, parts for both birch bark and canvas canoes, baskets, tools, and medicines.

Despite its low representation in the boreal forest, black ash and its ecosystem is nevertheless an important component of the boreal forest. It, and even more so its relative, green ash, are a major part of Thunder Bay's urban forest.

Even though the species is regarded as intolerant of shade, several years ago, a black ash naturally appeared in the shade on the north side of my house and I later planted a common forest associate, wild ginger, beneath it. The ginger is still there and spreading but I eventually had to remove the ash because it was too close to my house. The loss of either or both ash species in the bush or our communities would be felt by all NWO inhabitants. Unfortunately their near-future loss is possible and maybe even probable.

Anyone who has been reading about and listening to public awareness activities and announcements by the MNRF, Forestry Canada, and local Emerald Ash Borer (EAB) Task Forces is aware of the potential for loss of ash trees to this invasive insect. The EAB has reached Thunder Bay and Duluth, Minnesota, in its march west across North America and is leaving a path of dead ash trees behind. Some communities are proactively managing to protect their valuable true-ash trees and some have been partially successful in doing so.

Mountain ash is not a true-ash tree and is not at risk.

## RED PINE — IT HAS POTENTIAL

Red pine (*Pinus resinosa)* is found scattered across the boreal forest "from southeastern Manitoba to Newfoundland" and also across the northern states from Minnesota to Vermont and New York. Despite its large range there is "little genetic variation" in the species.[18] A red pine that is grown from seed collected in the Ottawa Valley will probably grow well with good form in northwestern Ontario.

That is not necessarily the case for other tree species that may have large ranges but wide genetic variation. For example trees grown from seed collected from jack pine and black and white spruce in one part of Ontario will often have different growth characteristics and form from trees grown from seed collected elsewhere in Ontario. These species have adapted well to local climatic variation and moving them outside the climate to which they have adapted can result in poor tree performance. For that reason, years ago Ontario foresters researched the genetic variation within those species and mapped reasonably distinct boundaries across which seed is not exchanged for commercial forestry purposes.

Red pine's needles, which are ten centimetres or longer, are attached parallel to each other within bundles of two on the twigs, distinguishing it from white pine with its five needles per bundle. Jack pine also has two needles per bundle, but its needles are usually less than four centimetres long, sharp tipped, and spread apart from the bundle base. Red pine's ten-centimetre long, ovoid seed cones mature in September and shed seed shortly after.

The species is intolerant of shade and mature trees have thick, fire-resistant bark, as does white pine, which the species often associates with. Their crowns are also usually above those of other trees, placing them above the killing reach of some wildfires. Most of the naturally regenerated red pines found in the boreal forest are thus of fire origin and found in even-aged stands. The tree is relatively deep rooted and wind firm and does well on dryer sandy and rocky soils.

I spent a worrisome night while solo camping on an island in Northern Lights Lake several years ago. I chose to pitch my tent under a mature red and white pine stand. I went to sleep while looking through the open door admiring the reflection of stars on the mirror-like lake surface. During the middle of the night a severe thunder storm with extreme winds struck and I could hear trees falling all around. Come morning I again counted stars, my lucky ones this time, as much of the "wind-firm" stand now lay piled like jackstraws.

The species is relatively disease free, but three boreal mammals can cause severe damage, particularly to planted trees. Porcupines often kill or deform a number of young red pines in a stand by eating the bark from large portions of the trees' trunks, effectively starving portions of the trees above the damage. Snowshoe hares kill seedlings by browsing on their foliage and stems, but they also starve, and thus kill, saplings by eating their bark at the stem base. White-tailed deer browse the tops off seedlings and small saplings, thereby stunting their growth and causing stem deformity. The plantations that I have observed in recent years, however, do not appear to have been damaged significantly by any mammals.

Red pine is a relatively easy tree to manage in the forest, and over the past twenty-five years red pine has been extensively planted, mixed with white pine, as a small component of many industrial spruce plantations in NWO. My strolls during recent years in some of the oldest of those plantations have impressed me with the health and superior growth of both pines.

Dr. Willard Carmean, professor emeritus in the Faculty of Natural Resource Management at Lakehead University, has for years been a strong proponent of planting more red pine because of its many positive aspects, but mainly because of its superior growth performance in our area. According to a brochure published by the MNRF, pure stands of red pine, which were planted during 1951 near Thunder Bay, by 1988 demonstrated total volume growth that was more than twice as high as that in similarly spaced adjacent white spruce and black spruce plantations.[19] Perhaps Dr. Carmean is correct and we should be planting more red pine especially now with increasing evidence of a warming world climate.

# 5

# FIRE — NATURE'S RENEWAL METHOD IN THE NWO BOREAL FOREST

Not only is it necessary to have a basic knowledge of the individual tree species and their needs, it's important to examine how they relate to each other and their environment.

To help us do that we will begin by following a hypothetical, but common, northwestern Ontario (NWO) wildfire as it starts, develops, and ends.

Imagine that we have been experiencing a normal slow start to spring. It is just past mid-May, all the snow is gone, but we have since been experiencing days of cold, sunless weather and tree buds are just starting to swell. The forest floor beneath deciduous-tree stands is covered with dry dead grass, herbs, and fallen tree leaves. Swamps and even some small shallow lakes are covered with last year's dead and drying vegetation. Both recently logged and burned areas contain mixtures of dead grasses, herbs, and a variety of fine to coarse woody material. Mature and old-growth stands contain numerous small to extensive areas of dead-standing and downed trees, which are the result of recent extreme winds and insect depredation.

We did have a thunderstorm pass through last week but it carried only light rain. It's now the third week of May and we have finally had a couple of 20°C days and today started like it is going to be another warm one. The Ministry of Natural Resources and Forestry (MNRF) has asked for caution with fire from all travellers and anyone in and about the bush. The humidity is expected to be lower and the temperature may climb into the high twenties with a stiff southwest wind — all conditions that, at this time of year, could cause a fire to be troublesome.

At noon it is warmer and dryer than forecast and at 2:00 p.m. it is 31°C with 28 percent humidity and gale-force wind. Fire danger in the bush, regardless of whether it is spring or summer, is extreme with those weather conditions, but the amount of dry, fine fuels on all ground cover types now makes the bush potentially explosive.

A remote old-growth stand of large, declining old jack pine, smaller but equally old black spruce, younger balsam fir, and some large old trembling aspen, contains scattered large white birch trees. Many of the old pines, aspens, and birches are dead or dying from rot and many have fallen in wind storms. They have been replaced by dense pockets of young balsam firs that are eagerly making use of the new sunlight getting into the stand. The conversion to fir could continue for years; however, a hollow old birch was hit by lightning during last week's thunderstorm. This stand has been a fire waiting to happen and today is its day.

The strong wind has fanned smouldering embers in the rotting interior of the lightning-struck birch and an opening in its broken top is creating a natural chimney. The rapidly developing flames make a loud whooshing sound that increases in intensity as they carry burning bark and rotten wood high above the old stand.

A vigilant worker at a remote mining exploration camp downwind of the birch tree smells smoke and calls the MNRF fire centre. A spotter plane is sent to check out the location but upon becoming airborne the pilot sees the smoke and immediately radios for a water bomber to tackle a rapidly growing fire just a few kilometres southwest of the mining camp.

Before the water bomber can reach the location there are already several fires burning hundreds of metres apart downwind of the birch tree. The spotter calls for additional bombers and ground crews but everyone is aware that the task of controlling the fire has become virtually hopeless until the gale subsides. Unfortunately an updated forecast now predicts it will continue into tomorrow evening, when, as the wind drops, the forecast is for the humidity to rise and the air to cool moderately.

At 4:00 p.m. a firestorm is racing above the treetops, across swamps, and even some reed-filled small lakes toward the northeast faster than a human can walk and all persons within twenty kilometres downwind of the fire are being advised to leave as soon as possible as their safety is at risk. The mining-crew supervisor orders a helicopter and by nightfall all personnel have been evacuated to a safe location, but they have had to leave most of their equipment in the fire's path.

The forecast proves accurate and late on the second day of the fire the MNRF deploys several attack crews to attempt controlling the fire's northwest and southeast flanks. There are no roads within thirty kilometres so it will be several days before bulldozers can reach the fire to create firebreaks.

The evening's forecast for the third day predicts continuing more favourable humidity and temperatures and only light, northerly wind that won't rise until mid-afternoon. The fire boss decides to create several small back-fires to burn fuel ahead of the fire's front. Taking advantage of the positions of several small lakes, a connecting creek and swamp, and a few aspen stands east of them he develops a plan. The following noon, using a helicopter with fire-starting equipment, he ignites areas that burn to close gaps in the natural fire breaks. By day five the weather is again warming but by now tree leaves and herbs are emerging, and these developments raise the moisture content of fine fuels and humidity near the ground. Wind velocity remains low and fire crews successfully hold the fire's perimeter for ten days until rain finally arrives and continues unabated for three whole days. After three weeks of hard work at digging out hotspots the MNRF declares the fire out.

Our hypothetical fire has consumed almost a hundred square kilometres of primarily old growth jack pine/black spruce stands with some interspersed younger stands on a variety of soils and terrain.

Now let's begin looking at how the stand in which the fire originated could react to the wildfire that occurred at this time of year and the prevailing conditions so far described. Fires occurring at different times and under different conditions would stimulate different responses in this and other stands.

## DIFFERENT SPECIES OF TREES REACT DIFFERENTLY TO FIRES

We start our look at stand regeneration following our hypothetical fire by re-examining the stand in which the fire began. It was a stand of large, old jack pines, smaller black spruces and balsam firs, and some large trembling aspens and white birches;

most of the old pines and poplars were dead or dying and many had fallen in wind storms. The dead-standing and fallen trees had been replaced by dense pockets of small balsam firs.

All the mentioned species, except balsam fir, are capable of regenerating following an early spring wildfire. However, the number of trees and their age, health, and seed load, along with soil conditions and subsequent weather patterns, can favour one or more species over the others.

Balsam fir has no seed-bearing cones before late August so it will be temporarily absent from most of the new stand, except near the edges of the burn where seed from adjacent unburned trees may find their way onto the burned ground.

The fire opened the jack pines' cones, but the trees were old and most had died. Even those still alive before the fire had thin crowns and few seed cones remaining. The volume of jack pine seed that falls after the fire is thus light. A small number of pine seeds will germinate and initially find enough space to grow.

The fire didn't occur because of drought; the soil was still damp and trees had adequate moisture. The flames raced through the stand consuming only the dryer, finer fuels and left scorched dead foliage high in the coniferous-tree crowns. Because the fire passed quickly the soil didn't heat up enough to kill the underground trembling aspen clones. They will soon erupt, pushing vigorous stems into the full sunlight. Aspen will quickly stake its claim to most of the growing space over the burned soil before most of the pine seedlings can get started, making it difficult for the pines to reach the full sunlight that they require. Those that do, however, will grow to maximum sizes for the species.

Most of the black spruce in the old stand would have been in the understory and would have had few seed-bearing cones;

however, a few larger trees that had been released by the early death of some shading pines and aspens will probably have significant seed loads. Black spruces will regenerate on at least equal terms with the pines, but they, too, will suffer in the deepening shade beneath the rapidly developing aspen canopy. Only a very small number lucky enough to begin their lives in the few openings among the aspen and pine, or under pine, will survive into the maturing stand.

Unfortunately for the birches they were past their best-before date and their ability to send up stump sprouts had dramatically declined. Some of the larger trees may temporarily survive the fire and next fall they will drop their seeds, while other seeds will be distributed from unburned islands within the burn's perimeter. The new stand will have scattered birch trees and the combination of all tree species will be what foresters call a "mixedwood" stand, but with a predominance of aspen.

Within the new stand, soil texture, structure, and fertility variation, and before-fire species distribution will influence the way in which species mix and compete.

If the fire had been as little as two weeks later after green-up, but after continuing high temperatures and low humidity, the developing new stand would have been much different.

## MID- TO LATE-SUMMER DROUGHT HAS DIFFERENT IMPACT ON POST-FIRE STAND

We are still looking at the stand that was burned in our hypothetical spring fire. Now we examine what might be expected if our fire had occurred during a mid- to late-summer drought.

The burn pattern would probably be different; deciduous stands would probably not have burned, nor would some of the grassy fens and bogs dependent on the degree of drought. Fire would less likely burn developing green bulrushes, reeds, and wild rice that are standing above water surfaces. These natural fire breaks would have protected some portions of coniferous stands that would not have been protected in the spring fire. However, a wind shift or increased velocity could cause some portions to burn.

Middle-aged, fire-origin stands of pine and spruce that, because of the scarcity of ladder fuels in them, would less likely have burned in the spring fire, would readily burn in this summer-drought fire.

Within most stands that did burn the fire would have been more intense and longer lasting. As a result, it would have consumed larger fuels and much organic soil, and heated the mineral-soil surface. All of our hypothetical old stand would probably have been consumed, except for some standing, larger diameter tree trunks.

All balsam firs would be consumed but regardless they would still have had no seed-bearing cones until late August, so this species will again be temporarily absent from most of the new stand except near its unburned edges.

The dead and moribund jack pines, because they had thin crowns, burned less intensely and may still provide a light seed fall, but the fire created an excellent seedbed for the species. More will survive and have optimal growing conditions relatively free from other plant competition.

The mineral soil was heated enough to kill many of the growth tips on near-surface underground aspen clones. However, some clones where the fire was less intense will sprout but with reduced vigour. The major part of the developing new stand will not likely be dominated by aspen, thus giving other species a better opportunity to regenerate and prosper.

Some taller black spruces will retain dense clumps of scorched cones at their tops and the cones will contain live seed. The species will again regenerate with the pine and will also survive better. Indeed because of the wide spacing of regenerating pines and aspens, the spruces will find numerous locations where they can develop on more equal terms with them, or even predominate in full sunshine.

All birch trees in the old stand would have been killed, but some seed will still enter from the burn perimeter and unburned islands within the burn. Because of less competition from aspens, the new stand may have more birches than did the old stand, but it will probably consist mostly of jack pine and black spruce with variable mixtures of aspen and birch and, after a late-August fire, some fir. All ground area will have tree cover but those trees will be wider spaced with larger crowns than those of the dense aspen that regenerated after the spring fire. The new stand will also be a mixedwood stand, but this time with more coniferous than deciduous trees. If our burned stand had been younger, the greater seed supply of pine and spruce would have enabled a denser, almost pure stand of those species after the fire.

Within our old stand there was undoubtedly some variation in the underlying soil, species mixture, tree ages and heights, terrain, and distance to the edge of the burn. These factors combined with prevailing weather into freeze-up, will collectively or separately influence the establishment success of one species over others at different locations within the new stand.

Over six decades at work and leisure I have enjoyed visiting freshly burned, harvested, and other disturbed areas to try to determine the contents of pre-disturbance stands, as well as speculating on the possible futures of the regenerating new stands. During later years, I have revisited many of those areas to check stand content and continue speculation on stand development. Year by year my speculation has improved, but I am still often surprised, awed, and humbled by the sometimes large impacts of previously undetected small and subtle variations.

## NO FIRES AFFECTS FOREST HEALTH

Sometimes a small, apparently insignificant action can alter forever the future of hundreds of square kilometres of forest. By extinguishing a lightning-strike fire in its infancy we often ensure the continuing presence of old stands for longer than is normal. Inevitably in the absence of fire old stands eventually convert from an original even-aged condition of light-loving pioneer trees to uneven-aged, late-cycle, shade-tolerant species.

To better appreciate this phenomenon let's take yet another look at the same old stand that I have been describing but this time it will not be burned. The birch tree that had been struck by lightning was discovered and reported by a timber cruiser several days later and she quickly felled it and extinguished its smouldering contents. Because of her action the spring firestorm did not occur and our hypothetical stand and its neighbours remain to face a different human-determined future.

Many years before our lightning strike and the cruiser's fateful action, several insect species began promoting tree-species conversion in the stand. Sawyer beetles and horntails were feeding on the dying and dead jack pines, hastening the death of some of them. Also an epidemic of forest tent caterpillars hastened the deaths of some of the aged aspens and birches and weakened most others. Now the stand is feeling the early effects of a developing epidemic of spruce budworms.

Stands like this one are not commonly vulnerable to the budworm because of the low presence of mature balsam fir. Nevertheless, some adjacent, similar stands on similar soils experienced a stand-replacing windthrow some forty years ago. Those stands then converted from the existing species to almost pure stands of fir with scattered white spruce. That conversion was made possible by the thousands-per-hectare of tiny twenty-year-old fir and fewer white spruce seedlings waiting in the shade on the forest floor in such old-growth stands. Because they were already there waiting for full sunlight, they got the jump on other species. Jack pines and black spruces had to wait for heat to open their cones and release their seed. Few of the seed found receptive seedbeds on the thicker, live moss and unrotted and rotting debris (duff and humus) of trees and other plants. By that time the firs and white spruces were already ahead of any other boreal trees, and as their elongating branches met they crowded out practically all other trees.

Those stands are now being stripped of foliage and support high numbers of budworms. Last summer a budworm-moth flight invaded our old stand and now their worms (larvae) are attacking the black spruces and balsam firs. Some of the shaded, thin-crowned spruces and firs are already leafless and, if fed on for a couple of more years, some will later die. Because of the combination of old age, insect and fungal attack, and wind damage, within as few as five years most of the old trees of all species except the larger, full-crowned black spruce and an occasional white birch and aspen will probably be dead.

Balsam fir will out-compete most other tree species and dominate the new stand except for the scattered old survivors. Aspens will emerge from root clones in the full sunlight of any larger stand openings created by the wind felling several trees together. Increasing numbers of shade-tolerant white spruce will gradually emerge as rodents and birds scatter cones and seeds from distant individuals. Light-loving jack pine will be absent from the new stand because of its inability to compete with fir, white spruce, and aspen under these conditions.

This stand will still be what some consider "boreal old growth." In reality, however, few, if any, of the trees will actually be beyond middle age. Like much of today's old-growth boreal forest, it continues to exist because of human intervention in the natural processes.

# 6

# HARVESTING THE BOREAL FOREST

There is a common belief that clear-cut timber harvesting is wrong in the boreal forest. Is it?

Of course, there are many degrees of clear-cutting, and there are alternative silvicultural systems to clear-cutting. The degrees range between felling and removing all standing trees and leaving patches, or numbers, of standing trees for a variety of reasons. Those reasons include habitat preservation, aesthetics, and riparian protection, among others. Additionally, clear-cutting can be done in two or more stand entries, such as in a final harvest after thinnings that are designed to improve the total yield of commercial products from the land area.

The shelterwood harvesting system is a variation of the clear-cut system. Both systems promote future, even-aged stands. In this system, there are two or more harvests, the first of which leaves sufficient mature trees of the desired species to adequately seed the harvested ground. The reserved trees can be evenly distributed, or in strips, or patches. When the harvested ground is adequately stocked with the desired species, the reserved trees may then be harvested. The time between the initial and final harvest affects the degree of uniformity in future tree ages.

Because of shade cast by the reserved trees, this system works poorly for jack pine and trembling aspen. Many of the jack pines that regenerate in harvested strips tend to come from trees that shed their seed without high heat; such stands tend to exhibit genetic degradation. The trees are less able to regenerate after fire if their cones no longer contain seed. Coincidental with that, jack pines that regenerate in strips tend to show poorer form.

The next stand also tends to have higher content of shade-tolerant species, such as balsam fir. Black spruces left as the shelter trees experience variable windthrow, severe as individual seed trees, and less severe when they are left in patches or strips.

The selection silviculture system is regarded by many opponents of clear-cutting as the gold standard of the available alternatives. In that system, stands are managed to create and/or maintain an uneven-aged structure. Harvest entries can be done after regular or variable time periods. Entries are designed to remove less desirable or diseased trees, enhance the growing opportunities of reserved trees to increase their future commercial value, and encourage regeneration of desired species. It works best in stands of shade-tolerant tree species and not at all for shade-intolerant species. It tends to work well in tolerant hardwood stands, such as those of southern Ontario.

Clear-cutting and even-aged stand management is the predominant system practised from the British Columbia rain forest across the boreal forest to Newfoundland. Clear-cutting can be used successfully in almost any forest if due care is used to ensure the regeneration of desired species. That said, I acknowledge that there are other values than regeneration to be considered. My practical experience has been limited to a relatively small area of the Canadian boreal forest. Knowing that climate, soil, and genetics among, and within, species vary across the boreal forest, I consider myself unqualified to suggest how other forests should, or should not, be managed.

## CLEAR-CUTTING IS PRACTISED WORLDWIDE

Take a world tour via Google Earth and fly over countries such as Finland, Sweden, Russia, the United States, Brazil, and Indonesia, to name a few, and you will see clear-cuts of a wide range of sizes. Finland probably has the smallest clear-cuts. "Clear-cut areas in Finland are smallest in family forests in the south, 1.2 hectares on average. The average for the whole country is only 1.5 hectares."[1] Their small size is partly the result of response to criticism, but it is also an attempt to follow natural patterns, "… clear-cutting areas are not defined arbitrarily but according to guidelines given by the nature."[2]

In other countries, however, the sizes of the clear-cuts are often seventy or more hectares and are sometimes contiguous or have narrow, often linear, corridors between them. The corridors tend to resemble the stream reserves and less-linear wildlife corridors that are left in Canadian clear-cuts. Indeed, when viewed from the air or via satellite (such views are available on a number of sites, including Google Earth), such clear-cuts often have straight, sharply defined edges, edges that appear to follow ownership boundaries.

## THERE ARE DIFFERENT OPINIONS

The natural forest pattern found in the boreal forest of north-western Ontario (NWO) is a jigsaw puzzle of tree stands. Pieces of the puzzle are tens of square kilometres of contiguous stands of trees, all of the same age, sometimes of a single species, but more often of two species. Stands with more than four tree species are rare. I maintain that if we are going to continue living close to, and in, that forest, while maintaining its natural structure, flora, and fauna, our choices are limited.

We can allow human-caused and naturally occurring wild-fires to renew the landscape as they have done for thousands of years. However, we would then have to move to, and concentrate development in, natural or human-made safe zones. That would require a massive redesign and relocation of homes and infrastructure, and we would still have to accept the occasional severe loss of life and property — loss perhaps even more severe than that which Fort McMurray experienced in 2016.

I am convinced that the only practical alternative to that disturbing scenario is to remain where we are and work with nature. That will require us to prevent and control wildfires where practically possible, and utilize and renew the forest with methods that maintain a landscape that is similar to that created by wildfire.

I have some questions that I believe we should all ponder with open minds.

- Is our reasoning rational when we see something that, to us, appears ugly?
- Are some of us selfishly abandoning objective reasoning to achieve our own ends?

- Would we be less satisfied with the consequences of alternative harvesting techniques than we are with the consequences of clear-cutting?
- Are the things that disturb us most about harvesting based on past careless practices that are currently rare or discontinued?
- Are our concerns really supported by the best available current knowledge or are we cherry-picking information that supports our arguments?
- How do the impacts of clear-cutting differ from those of wildfire?
- Are those impacts good or bad, and against which standards are the impacts judged?
- What will happen to plant communities and wildlife habitat if we ban clear-cutting in favour of partial (selection) harvesting?
- Will wildfires burn more intensely in selection-harvested stands?
- If so, what will be the new plant and wildlife succession?
- What will happen to existing development and homes if we choose wildfire over harvesting?
- Which will have the greater impact on climate change — wildfire, selection harvesting, clear-cutting, or no harvesting?
- How many of us are focusing on the twigs in other's eyes and ignoring the logs in our own eyes?
- Is our concern about current harvesting techniques distracting us from taking action on some of our own more environmentally destructive practices?

I offer the following quotes from one of the Ministry of Natural Resources and Forestry's (MNRF) silviculture guide books that was based on the available research and experience of practitioners:

> The Boreal Forest is disturbance driven and dominated by species that are adapted to these environmental conditions. Fire, wind, and insects often affect extensive areas of forest, initiating regeneration. These areas normally regenerate to even-aged stands. Clearcutting, involving a range of cutblock sizes, creates landscape patterns which can approximate those created by natural disturbances.[3]

> The selection silvicultural system is not suited to management of intolerant species such as jack pine or aspen. It is also not applicable in even-aged black spruce stands (Groot 1994).[4]

> Applying the selection system on upland black spruce sites can lead to a species shift towards balsam fir and white cedar (Johnston and Smith 1983). The selection system may be biologically appropriate for the management of uneven-aged and/or uneven-sized low-land black spruce ecosystems (Groot 1994). Although research trials exist, this technique has not been applied operationally in Ontario.[5]

My experience in the NWO boreal forest has confirmed for me that the silviculture guides that have been produced by the MNRF are well founded and based on evidence, and I look forward to future revisions as new research and experience suggest changes are required. I have detailed how the variety of trees and some other plants respond to common disturbances, how they interact and follow or eliminate each other in the natural forest succession. The unique roles of different species as I have described them will explain the following paragraphs.

I am afraid that if we abandon clear-cut harvesting in NWO, the only way to maintain the natural forest succession will be to allow uncontrolled wildfire across the landscape. My experience, lifelong observation of the boreal forest, and formal study lead me to believe that a selection-harvested forest will follow a successional pathway that we will not like. That forest will have fewer jack pine, black spruce, trembling aspen, white birch, and eastern larch and more balsam fir and white spruce. We can then expect more insect epidemics, tree diseases, wind damage, and possibly unwelcome changes in wildlife species and populations. Even worse, we will be inviting even more intense fires in a forest that, because of our actions, has few if any trees that are capable of regenerating following the fires. There will be low chance for the forest to recover to its historic condition.

Of course, not everyone agrees with me and that is the way it should be. If we all agree on something, we may be agreeing on a big mistake. However, I invite, no, I plead with my readers to pull back from firmly held pro, or con, positions that you may have about even-aged forest management and clear-cutting; consider with open minds what I and others are saying, and think about our knowledge and objectivity. The future management of the boreal forest will be based on our collective judgment.

Current public attitudes toward forest management practices are revealed in an article the *Toronto Star* published in December 2014. According to that article, the results of a survey by OraclePoll that sampled the responses of Ontarians to the draft of a forest management guide being developed by the MNRF showed that "92 percent oppose a Ministry of Natural Resources proposal for larger clear-cuts. Environmental groups such as the Wildlands League, Earthroots, the Federation of Ontario Naturalists, and First Nations leaders have long been among a growing list of groups opposed to these measures."[6]

I understand that the guide was being designed to help forest managers emulate natural disturbance patterns as they harvest the forest. Large disturbed areas within the forest appear to be part of the long-term habitat needs of forest-dwelling woodland caribou. That animal, like the boreal forest tree species it dwells among, has evolved in response to huge wildfires and may be dependent on large even-aged stands as part of its habitat. Moose, whitetail deer, and consequently wolves, a major predator of caribou, thrive in a forest of smaller, fragmented harvests, or natural disturbances.

The draft guide proposed a mixture of small and large clear-cuts with up to 20 percent larger than 260 hectares in area. Approved management plans and subsequent harvesting plans would have to contain valid reasons for the size and location of each such harvest.

I am convinced that right and wrong are constructs of our human minds and that nature makes no such distinctions. The forest has evolved with resilience to eventually return to its historic structure. When we recognize that resilience, and accommodate it by utilizing the forest with methods that best enable the continuation of natural trends, we hasten the return to a natural forest structure. My formal studies and long experience observing the boreal forest tell me that clear-cut harvesting is the silvicultural technique that best enables those trends to continue in NWO's boreal forests. Larger cuts better represent the forest that natural forces created and better accommodate the wildlife species that have evolved in that forest.

I acknowledge the concerns of persons outside of, and within, the forestry community about the environmental impacts of clear-cutting. I believe that in most cases opponents to clear-cutting are basing their opposition on their own experiences, studies, and knowledge of others, and that they should be considered when making forest management decisions within current policy and legislation.

My purpose in writing this book is not to argue against the legitimacy of the concerns of others about forest management. Instead, I feel a responsibility to help ensure that the advantages of even-aged stand management and the clear-cut silvicultural system in the boreal forest of NWO are well understood by an engaged public.

## CAREFUL HARVESTING — MINIMIZING DIFFERENCES BETWEEN NATURAL AND MANAGED FORESTS

How do the impacts of clear-cutting differ from those of wildfire?

The impacts depend on variations within and among a complex array of factors on each different fire and cutover.

Included among those factors are weather, terrain, soil fertility and drainage, micro-climate, pre-disturbance forest conditions, timing of disturbance, and the degree of care exercised by the harvester. A full explanation of those variations and specific impacts would require discussion of volumes of scientific data beyond the scope of this book; however, some differences do stand out.

After a fire in a near-mature or older stand, many dead trees remain standing, a few for several years. Those trees, depending on species, age, time of the fire, and other factors, may bear seeds that regenerate the burned forest.

Tree regeneration on all soils is enhanced by fire, but enhancement is more pronounced on richer soils than on dry or wet, less-fertile soils because of the tendency for richer soils to support more prolific and aggressively competing vegetation.

When wind blows standing dead trees in a burn, their roots flex, and this motion creates cracks in any remaining organic mat. This action gives black spruce in particular better germination and survival chances.

After clear-cutting, fewer standing trees remain than after fire; however, today's harvest prescriptions require that some standing trees be left for wildlife roosting and nesting use. Regeneration of the ground from the seed in those trees is less effective than after fire, especially on richer soils, because ground disturbance by harvesting machinery stimulates, rather than depresses, growth of grasses and hardwood shrubs. Additionally, those trees, because of the wind-sail effect of their living crowns, suffer earlier windthrow or breakage. To prevent toppling or breakage and prolong their usefulness, the tops are cut off some trees so that they more nearly resemble natural old snags.

Trees beside lakes and streams are usually reserved during harvesting, but increasingly, scientific study is finding that practice is unnecessary and may even have negative consequences for some species, such as beaver. Fire doesn't recognize any need or consequence and will carelessly burn to the water's edge.

Possibly the most significant consequence of fire to the future forest is the near elimination of shade-tolerant species in favour of those that require full sunlight. Clear-cutting, on the other hand, seldom eliminates any species and the new stand will usually contain a few more shade-tolerant trees than it would if it had developed following fire.

I have noticed that ground lichens and reindeer mosses tend to remain relatively healthy on coarser soils and exposed rock on some harvested areas, whereas they are completely eliminated for up to thirty years by wildfire.

Harvesting machinery tends to crush most small trees and other plants that existed prior to the harvest. Nevertheless, trees that are regenerated — either naturally or artificially — after the harvest are the same as those regenerated by fire, particularly on poorer soils. On richer soils, subsequent stand renewal and plant-competition-control treatments help ensure that the developing new forest also has fire-encouraged species.

Severe drought-enabled fires remove almost all organic soil cover, may significantly alter soil chemistry, and, by eliminating most plants, render soils susceptible to erosion. Most of the organic mat and plant roots remain after clear-cutting, and when combined with today's improved road construction and machinery, reduce the chance of soil erosion.

Forest managers can, with planning and care, assist the resilient forest that develops on harvested areas to gradually converge toward a more fire-type forest, thereby reducing earlier differences between the two.

## PARTIAL-STAND HARVESTING — CHANGING THE BOREAL FOREST AWAY FROM ITS NATURAL PATH

Would we be more, or less, satisfied with the consequences of partial-stand harvesting than we currently are with those resulting from clear-cutting?

Would we be more, or less, satisfied with the consequences of no timber harvesting?

I suspect that we would be less satisfied with both partial-stand harvesting and no harvesting. I present this viewpoint and ask the reader to ponder this subject before continuing. Before we demonize the legitimate, well-tested clear-cut system, which in my experience-based opinion is currently working, we need to objectively recognize and question our biases, motives, values, and hopes for the future.

I have described how the boreal forest of NWO evolved with wildfire. After a century of clear-cut harvesting there are still lots of fire-origin stands with variable mixtures of jack pine, black spruce, trembling aspen, and white birch (let's call it PSAB for short). Each group of PSAB stands that originated from the same fire contains trees all of the same age. Large numbers of such stands collectively cover hundreds of square kilometres. They often abut other such collections of stands that originated after different fires, and each collection has a different single age.

In my opinion, Ontario has wisely chosen to maintain that

natural tree-species and landscape pattern to sustain the forest and its inhabitants. How can that be done best?

I have expressed doubt that we want to depend on wild-fires. However, regardless of our best efforts, wildfires will continue to replace a significant area of forest over the long term.

All four PSAB species regenerate and grow well in full sun-light. Full sunlight is not available unless the previous forest has been virtually eliminated. Three of the species regenerate and survive poorly under the shade of larger trees in partially harvested stands, and the fourth does not thrive there. The only alternative we have to achieve Ontario's objective, that I can see, is to clear-cut the forest in fire-type, large areas and patterns so that together wildfire and clear-cutting maintain the natural species and stand patterns.

Most mature tree stands in the NWO boreal forest, regardless of stand origin, have tiny, shade-tolerant balsam fir seedlings happily existing in the shade on the forest floor. As soon as an opening is created among the treetops these seedlings begin rapid growth, and as the opening size is increased the firs have usually already grown enough to shade the ground, preventing any PSAB seedlings from regenerating or thriving. On most soils white spruce, another shade-tolerant seedling, will join the fir. Over time, in the absence of fire or clear-cut harvesting, these two species dominate most stands.

Partial-stand harvesting, by creating small openings, creates optimal conditions for fir and white spruce, giving them the advantage over the PSAB species. So let's try to picture a stand that is being partially harvested.

During partial harvest we remove the prescribed crop trees leaving remaining trees to grow. At some prescribed future year we re-enter and remove another portion while in the meantime the main species that replace those being harvested are fir and white spruce.

The fir and white spruce can be removed but the regenerating conditions remain the same. Only black spruce among the PSAB species readily regenerates in the partial shade of the PSAB group, but at little advantage as it grows poorly in the shade of the larger trees. As we remove their neighbours, the shallow-rooted black spruce are increasingly susceptible to being blown down. I anticipate that within twenty or fewer years of partial-stand harvesting, we will all agree that we have an environmentally and economically degraded stand that is highly combustible and with limited ability to regenerate to trees after fire.

Let's not make that mistake. Let's continue to provide forest managers with the professional scope to prescribe and use silvicultural systems that they know will work and have a better chance of satisfying our future wants and needs.

## PARTIAL-STAND HARVESTING — IT'S RISKY

While growing up in a logging town I was appalled at "indiscriminate clear-cutting" across a lake from my home. It was unsightly and I believed that the forest would never grow back to its former vigour. It did grow back, but in my self-interested view I continued to see it as less desirable than it had been before it had been logged. In Chapter 2 I said that I was a logger who clear-cut the forest and wrote that as a forest manager I prescribed clear-cutting, and if still practising today, would still do so. Why did I change my mind, or did I compromise my principles?

One spring during my studies, while waiting for a summer job with Parks Canada to begin, I took a job cutting pulpwood, hence the clear-cut. Yes, I did feel guilty at the time but I needed the money and I was able to rationalize my situation; I was educating myself to do something about it later. I had visions that, after graduating, I was going to change the way timber harvesting was done. I hoped to return to Anglo (Nfld.) Development Company Ltd. (AND) and make changes from within.

I joined AND, by then Price (Nfld.) Pulp & Paper Limited, immediately after graduation and made it known that I didn't like clear-cutting. Maybe that was why my boss and mentor, Frank Hayward, decided I should eventually re-measure, analyze, and report on the old stand management trials mentioned in Chapter 2. Today I realize that he knew an objective analysis of the trials would show that clear-cutting was necessary for the long-term health of the forest.

The trials and subsequent experiences clearly demonstrated to me that, if we are going to maintain the primary boreal forest's large-area, even-aged structure of multiple, single-aged stands, then large-area, multiple-stand removal is necessary. Nature's way to do that is with fire but the only alternative that I could see for me, if I was going to prescribe harvesting, was to prescribe clear-cut harvesting.

Large-scale, diameter-limit harvests were initiated in 1946 in each of the four logging divisions of AND. My concluding measurements of 1966 showed that there had been no net commercial yield increase and the stands were trending toward balsam fir dominance. A diameter-limit harvest is one in which all trees above or below a certain diameter, as measured at breast height (1.37 metres), are harvested.

Measurements of other commercial thinnings, started in the 1920s and 1930s using various "rules-of-thumb," at Pamehoc Lake and Gill's Pond, Newfoundland, ended in the 1960s with the same results. Over time the clear-cut stands that were part of each trial had more black spruce and later studies of mechanically harvested clear-cuts showed fewer firs and more black spruce.

A winter manual selection harvest in a black spruce stand, which was designed and managed by me in the 1970s, was not followed to conclusion after I left in 1978. Frequent inspection by me then revealed low individual tree response, increased wind damage, and almost no regeneration of any tree species.

There was a significant increase in merchantable volume for pre-commercially thinned (spaced) stands.[1]

Those studies were in Newfoundland where jack pine was absent and winds at that time were stronger than those that prevailed in NWO. Our reports were not published or officially peer reviewed but I am so far unaware of any published research that contradicts our observations.

Meanwhile, Abitibi Power and Paper studied various harvesting patterns on its private land northwest of Thunder Bay, Ontario. Two of three immature fire stands of jack pine and black spruce, which were partially cut to remove the jack pine and leave the black spruce, blew down shortly after harvest. Other partial-cut trials have trended to multi-aged, predominantly fir stands.

A seldom discussed impact of partial-stand harvesting is the need for the construction of a much higher density of access roads to extract the timber and greater harvested area to recover the same timber volume. Other impacts include increased habitat for some species such as marten, but decreased habitat for others such as moose.

I came to realize that partial harvesting of even-aged stands in the boreal forest is risky business. Additionally it is apparent to me that partial stand harvesting negatively impacts forest health. It builds a forest that is vulnerable to larger and more intense wildfires meanwhile increasing vulnerability to insects and disease and decreasing total-forest biodiversity.

## PARTIAL-STAND HARVESTING — A MORE DETAILED LOOK

Let's begin by examining how boreal trees collectively accommodate fire and other disturbances. We go back to how the boreal forest probably behaved and looked before Europeans arrived in our area.

Trees in naturally regenerated stands tend to be semi-randomly distributed, resulting in variation of distance from each other and a tendency toward clumping. That means that, in some parts of a stand, trees have more space to grow than they need and in other parts trees have less space than needed. This is the result of variation in natural factors such as soil disturbance, seedbed conditions, seed availability, wind direction and velocity at time of seed distribution, aspen-root distribution, and many others.

Of course trees that have more space than needed (lower stand density) grow to become larger than do those in parts of the stand that have higher density. Depending on species, that can result in trees that look dramatically different. A jack pine, white birch, or white spruce will concentrate more of its growth into larger, widely spreading branches and larger trunk diameter at some expense to tree height. On the other hand, black spruces tend to maintain a narrow crown with small branches and similar height regardless of stand density.

As time passes other trees may grow into some of the less-dense parts of the stand. During the first ten years or so those new trees can be any species depending on seed supply, seed bed conditions, and other plant competition. After the widely spaced trees start to spread their branches and create shade, any new trees will most likely be shade-tolerant species, such as balsam fir and white spruce, but could also be black spruce. We refer to those latter trees as "ingrowth." In denser parts of the stand, because of each tree's different access to water and nutrients and its genetic traits, some trees lag in growth and become suppressed, and if they are shade intolerant they probably die. We refer to that as "natural thinning."

After the branches (crowns) of all trees have met (crown closure), the ground becomes heavily shaded. The degree of shade depends to some extent on tree species, and whether they are coniferous or deciduous. Generally boreal deciduous species allow more light penetration of their crowns than do coniferous species. With crown closure, particularly in coniferous stands, the forest floor vegetation moves through a change of species and vegetation grouping. There is a general transition away from a preponderance of herbs and grasses toward more mosses and ferns or small woody shrubs, and on some soils under deciduous trees and widely spaced jack pines, to hardwood shrubs.

Beneath deciduous stands balsam fir and, less frequently, white spruce and occasional black spruce invade and continue to grow but at a slower rate than in full sunlight. All three of those species are capable of periodically responding positively to increases in light caused by loss of neighbours, only to slow down again as other neighbours encroach upon their new-found space.

As coniferous stands mature there is little ingrowth of deciduous trees, but beneath jack pine's thinner foliage, semi-shade-tolerant deciduous shrubs are common. Black spruce has denser foliage and allows less light penetration to the ground. Under a closed canopy of black spruce there will usually be fewer herbs or large shrubs, and the ground will almost always be covered by a variety of mosses, from the fine feather mosses on dryer soils to the coarse and absorbent sphagnum mosses on wet ground.

## ONE STAND TYPE — WATCHING IT START AND GROW

Let's follow the life of a stand of jack pine and black spruce on a moderately fertile, moist, upland soil. In 1980, when I arrived in Thunder Bay, the forest along the Spruce River Road (Highways 527 and 811) from Thunder Bay to Garden Lake was primarily of this type and much of it still is.

The majority of those stands originated following fires, and between 1900 and 1980, several large fires combined to remove thousands of square kilometres of mature forest. In 1980 one fire alone, TB-46 between Garden and Obonga Lakes, removed over twelve hundred square kilometres of mostly mature and old-growth (over-mature) jack pine and black spruce stands. Several square kilometres of that burn have since re-burned.

These types of stands can be predominantly either spruce or pine and have various mixtures of trembling aspen and white birch, but for simplicity we will limit discussion to those having 20 percent or less of the stand as aspen or birch. On the more fertile soils, grasses, herbs, hardwood shrubs, and often trembling aspen quickly populate the ground after most whole-stand-removing disturbances other than fire. After large drought-assisted hot fires most plants, other than jack pine and black spruce, have no seed remaining to populate the area or those that do are suppressed enough such that jack pine and black spruce can outcompete them. Low numbers of aspen and birch seldom interfere with the pine's and spruce's development and in some cases may even enhance their form and volume.

Earlier I described how the various tree species get started but now let's consider how they grow as a community. We have discussed how non-tree vegetation impacts tree establishment and early survival, but as stands develop roles reverse and trees impact what other vegetation inhabits a stand.

On some parts of our chosen stand widely spaced pines are the only trees, so let's call those areas the understocked portion of the stand. Other parts of the stand will have a mixture of pine and spruce and some aspen and birch, with most trees having optimal growing space. We will call those areas the well-stocked portion of our stand. In the remainder of the stand the four species are tightly spaced together, thus giving each tree less than ample room to grow. We will call that the fully stocked portion of the stand. Of course in any stand in the real forest any one or two of understocked, well-stocked, and fully stocked portions may be absent or form a greater or lesser percentage of the whole.

As the years pass each of the three described parts of the stand develops differently. The understocked jack pines may be repeatedly attacked by insects such as the white pine weevil and the pine shoot moth. Both can cause some trees to periodically stop growing vertically, develop large lateral branches and, in some cases, multiple main stems. Grasses and hardwood shrubs predominate on the forest floor.

Meanwhile in the well-stocked sections trees grow upward and outward until their branches meet, at which point lower branches begin to die from lack of light but higher branches continue to grow until they too meet and eventually decline. Depending on steepness, direction, and position of slope relative to the prevailing winds, height growth continues until maximum height inherent in each species and soil potential is reached. Tree trunk diameters are smaller than those in the understocked portion but the stems are straighter and have less taper up to their living crowns. The forest floor will have a collection of mossy patches intermixed with herbs and grasses and a few hardwood shrubs.

In the fully stocked portion tree branches meet earlier and the few faster growing individual trees overtop some of the slower growing ones, which later die. The remaining trees continue to compete for space and develop narrow, thin crowns, with the jack pines and aspens overtopping the spruces and most of the birches gradually declining and disappearing.

## PARTIAL-STAND HARVESTING — RULE-OF-THUMB METHODS USUALLY DON'T WORK

Now let's explore different ways to partially harvest our candidate stand. It's important to realize that the three levels of stocking just explained are usually distributed in variable-sized patches throughout stands and gradually blend with each other.

There are several ways in which we can partially harvest an even- or single-aged stand. Each method is best performed when the stand is immature. Immature, reserved trees have more time to increase foliage, root areas, and stem growth before reaching old age and tree decline. However, even old trees of some species can increase growth when they obtain more growing space after being released from competition.

We begin by examining less effective and move to the more effective methods of partial-stand harvesting. I again caution you that my early experience studying partial-stand harvests was in spruce and fir stands in a windy environment. Here in NWO with its inclusion of jack pine and a calmer environment (at least it used to be) some results may differ.

A diameter-limit thinning, in which all trees above a specified stem diameter are removed, is based on the premise that the reserved smaller trees will respond with increased growth. The result, in my experience, was that the understocked areas were clear-cut because all the trees there exceeded the diameter limit. The well-stocked portions varied between having too many and too few trees being removed because of variations in spacing and minor clumping. The fully stocked portions had almost no trees that exceeded the minimum diameter and therefore experienced no effective thinning.

Many reserved trees in the former well-stocked and fully stocked areas blew down shortly after the thinning because wind could penetrate the stand from the openings created in the understocked portions.

My report concluded that, after two decades, there was no net increase in merchantable volume. Some reserved trees showed increased growth while others were lost to windthrow or still had inadequate growing space to respond. There was a burst of merchantable volume growth on some sections but it was almost all on small, clear-cut, understocked portions of the stand that, before harvest, had small fir seedling and sapling ingrowth. They rapidly responded in full light and within twenty years reached merchantable diameters and effectively converted the stand from predominantly spruce to predominantly fir. I concluded that, for diameter-limit harvesting to work, we need uniform spacing between trees. Uniform spacing of trees is rare in the natural boreal forest and the inclusion of jack pine does not change that condition.

The positive growth that did occur was almost exclusively on the clear-cut portions of the stand and indeed commercial (merchantable) volume growth was negative in the theoretically thinned portions.

During the 1960s Abitibi Power and Paper experimented with thinning from above (cutting the tallest trees) in immature mixed jack pine and black spruce stands at its Abitibi Woodlands Laboratory on its private lands near Raith, Ontario. The company theorized that harvesting the taller jack pines would release the shorter black spruces growing beneath them to respond to increased light and growing space. Within a few months of the treatment, however, two-thirds of the thinned area blew down but the remaining one-third matured into one of the best black spruce stands that I have ever seen. Again, however, merchantable-growth response to the thinning was negative.

Black spruces are always vulnerable to wind. During the 1990s Abitibi-Price carried out a commercial thinning from above in similar but older stands on its private land. This time nearly 100 percent of the reserved black spruce trees were wind-thrown during the following autumn. Those two trials have convinced me that thinning from above for black spruce in such stands is very risky business indeed, even in NWO.

We will next examine other possible approaches that, with careful management, can be more successful but still, in my opinion, have some negative consequences about which we need more knowledge to avoid. There are thousands of square kilometres of young plantations and seeded areas that, with good knowledge and careful thinning, we may be able to guide toward higher-value production.

## PARTIAL-STAND HARVESTING — CREATING A FOREST, ONE STAND AT A TIME

Let's now assume that, when viewed from the air, we want an almost uninterrupted forest of mixed species in all directions to the horizon. Okay then, let's try to create such a forest one stand at a time.

Some tree species are going to be difficult to manage over the long term in what will eventually become multi-aged (uneven-aged) stands. The most difficult species will probably be jack pine, white birch, and trembling aspen. Regardless, we will maintain those species as long as we practically can.

We will create conditions that encourage as many tree species as possible by enhancing seed production, shedding, and germination, plus seedling survival and growth for some trees even into old age.

Jack pine, birch, and aspen require significant cleared areas in order to regenerate and grow, so we will create those spaces by clear-cutting the numerous small, understocked areas of large trees. The cuts have to be large enough to provide for pine's requirements, but birch and aspen can capitalize on smaller clearings. Ground preparation and seeding or planting will enhance pine's success.

On the remainder of the stand we will retain a standing forest and nurture those trees that exhibit what we deem to be the most desirable features for our purposes. The pines are relatively wind-firm but short lived, so if the stand is young enough let's reserve some of the larger ones. By removing some of their closest neighbours we will provide more space for them to increase foliage area (their growth factory) and enhance future trunk growth.

We can also reserve a few larger black spruces; however, we will have to be careful not to expose them too quickly to strong winds. Since the species can respond repetitively to release from competition and has a longer life span, let's remove one neighbour at a time. This can be done over several years, allowing the spruces to gradually enlarge their crowns and strengthen their root networks. With care they can become relatively more wind-firm and eventually more valuable sawlogs.

As we harvest trees, creating small openings in the tree canopy, sunlight briefly reaches the ground and advanced-growth balsam fir with a few spruce, followed by a few more newly regenerating spruce, appear in the openings. Depending on the number and sizes of the advanced growth fir, the spruce may or may not get a chance because they generally do not get started until after some sunlight reaches the ground. If the fir get ahead of the spruce they will probably prevail. We can hold fir back, at some significant expense, by cutting larger ones or scraping the ground to remove smaller ones, and simultaneously enhancing spruce regeneration.

The few aspen in the stand are probably among the larger trees and they are wide-crowned and heavy. Harvesting them without causing severe damage to many desirable trees will be difficult. We will probably leave them to decline naturally, die, and gradually crumble, unless at some time they are in demand and then we can make calculated trade-offs.

The end result, after all pre-existing trees have died of old age or been harvested, will be a primarily coniferous stand of shade-tolerant balsam fir, followed by white spruce, and black spruce. Black spruce will gradually be nudged to the wetter and dryer, less-fertile locations. There will be some pockets of light-demanding (shade-intolerant) species on the small clear-cut and naturally de-pleted locations. Less common and light-demanding eastern larch, white cedar, and black ash will continue to find satisfactory niches primarily in created openings on the wetter soils.

## OPPOSING NATURAL SUCCESSION HAS ITS CONSEQUENCES

In our area there is little conclusive research into partial-stand harvesting such as that just described; however, there is at least one project in progress in mixed-coniferous and deciduous stands. Meanwhile we have to rely on informed experience, what we know from science of the needs of the different tree species, and research elsewhere.

All natural tree stands have variations in tree spacing so, in that regard, they are somewhat like the pine/spruce stand that we just discussed. To create our uninterrupted forest from horizon to horizon we will have to apply selection harvesting to stands with a variety of species mixtures, ages, densities, and structures. In any of those stands we will use the same strategy of clear-cutting understocked locations and relying on natural regeneration of aspen and birch. We will plant mixtures of pine and spruce to enhance species diversity in those areas. To ensure that the planted trees survive, in some locations, we will practise some method of aspen, grass, and shrub control.

Well and fully stocked portions of most stands will probably have some wide-crowned, deciduous trees. We will carefully harvest some deciduous trees while trying to minimize damage to their surrounding neighbours. Where possible we will retain a relatively equal distance between reserved trees to ensure all the ground is fully utilized. One objective that will trump all others will be to create a well-stocked stand, reserving those trees that have the best potential to meet the forest's and our perceived future needs.

As we harvest individual trees, or small groups of trees, we will create small openings, allowing more sunlight into the stands, thus enabling small advanced-growth seedlings and saplings to increase their growth. As the reserved trees' crowns widen and meet, sunlight will again be restricted, limiting growth of smaller fir and spruce. Each subsequent harvest entry into the stand will cause this cycle to repeat.

Numbers of advanced-growth fir and regenerating spruce will depend on seed source, soil fertility, and moisture, and the density of aggressive herbs and hardwood shrubs. Birch and aspen will regenerate in larger openings that we can create, if that is desired; however, we may want to convert some stands to mainly coniferous species. If we depend on natural regeneration, the primary species will again likely be balsam fir. We can change that in mature hardwood stands by planting spruce beneath the reserved trees and removing advanced-growth fir. White spruce is more shade tolerant than black spruce, and will have a better chance to gradually grow through the deciduous-tree canopy. However, both spruces will prevail as the hardwoods die or are harvested.

All the original trees in the stand will eventually be harvested or die. As this occurs, unless we create larger openings for deciduous trees to succeed, almost all stands will gradually transition to more coniferous species.

In my opinion almost all varieties of stands on average to good soils in the northwest, when harvested in this way, will transition primarily to balsam fir with scattered white and black spruce and small patches of pine, aspen, and birch. Stands on wet soils will continue to be dominated by black spruce. Stands on dryer, less-fertile soils will transition to mixtures of fir and black spruce with some jack pine and birch in poorly stocked locations.

If we manage all stands in this manner we may be able to develop a nearly uniform, closed forest from horizon to horizon; however, climate change, with its current and expected erratic behaviour, will be the elephant in the room. Mother Nature is passionless, she deals from a stacked deck, and she always plays by her rules.

## PARTIAL-STAND HARVESTING — BE PREPARED FOR MANY MORE ROADS

Let's acknowledge the elephant in the room before we start looking at the consequences of current and future forest management options. Recent erratic weather variations have already disrupted, and we can anticipate that future variations will continue to disrupt, whatever forest management strategies we decide to follow. Climate change is happening, and there is increasing public, scientific, business, and international political consensus that the rate of change will accelerate and the degree of change will intensify.

In NWO during recent years, we've observed the following:

- extremely high winds have levelled hundreds of square kilometres of forest;
- severe rain storms have caused unprecedented flooding and property damage;
- the number of wildfires and areas burned have been decreasing despite high losses during some years;
- in March and April of 2012 we experienced unprecedented summer weather followed by a return to winter conditions. During that period trees broke dormancy and then froze, resulting in coniferous-tree-needle desiccation and loss of the current-year's growth buds. The end result was severe weakening and eventual mortality of some trees of all ages in limited locations north, west, and south of Thunder Bay; and
- more southern species of fauna have moved into our area in increasing numbers with potential, but as yet unmeasured, impacts on the forest.

Can we expect more of these changes? I am afraid that we can, and with increasing intensity, while combining with changes we have not yet experienced. While acknowledging unknown consequences of climate change, however, let's look at more predictable consequences of partial-stand harvesting.

Regardless of which way our climate changes, successful partial-stand harvesting will require periodic stand-tending and harvesting to enable enhancement of reserved trees and maximum recovery of usable wood. Repeated stand entries will require a dense, high-quality, well-maintained network of access roads. The number of entries into a stand will depend on what types and numbers of treatments we find are necessary to recover volume and quality of wood equal to that recovered from current harvesting practices.

During the first rotation, or the time it takes trees to grow from seed germination to maturity, those repeated stand entries will require us to build and permanently maintain roads on two, and even three, times as much area as we do today. That means more ground lost from tree growth and logging activity will occur on up to three times as much area. For those of us who are concerned about expanding road networks and their real, or perceived, impacts on the natural environment, that spectre should give us serious pause.

Today, after harvesting and stand renewal are completed on an area, all but the main roads are usually permitted to deteriorate, and the forest to slowly re-invade them. Eventually, even when viewed from above, unmaintained roads become all but invisible. In partial-stand harvests, such as I have been describing, almost all roads will have to be maintained indefinitely to enable continuing stand re-entries.

The cost of road construction and maintenance in northern Ontario is one of the greatest handicaps we face when competing with those who use public roads while harvesting planted stands in warmer climates. Each additional stand entry we make will require additional costly moves of equipment and lower machine and manpower productivity. Add to that costs of stand tending to reduce ingrowth of lower valued species and one begins to understand that managing with partial-stand harvesting in our area could become less attractive to forest-industry investors.

If we are going to seriously consider converting a major part of our forest from its natural single-aged, pioneer-species stands into less common, multi-aged, late-succession-species stands, we need to be researching the consequences now. There will probably be consequences that we do not currently anticipate and have no means of measuring, because they could affect a single species or even occur on a forest-wide scale; some consequences will be negative and some positive.

## SELECTION HARVESTING — ITS EFFECTS ON THE BOREAL FOREST

What will a NWO forest that we develop using large-scale selection harvesting be like?

I have never seen and don't know of a classic selection-harvested stand in NWO. I base my opinion on the following experiences:

- I've reported on trials of different methods of thinning that were monitored over as long as forty-eight years;

- I've studied changes that occurred over three decades in several hundred study plots in single- and multi-aged stands following fire and different degrees of harvest on the island of Newfoundland; and
- over thirty-four years, I've observed several stands north of Thunder Bay, noticing changes in them.

I repeat that my work was not officially peer-reviewed or published in accredited scientific journals. I am not currently a licensed professional forester and never was a recognized expert in selection harvesting. With those caveats I offer my opinion.

The most obvious impact of wide-scale, selection harvesting in NWO would be a gradual decline of jack pine in any stands so managed. Individual jack pine trees can't regenerate or grow well in the shadow of other trees.

As the jack pine decline, if they are present, understory black spruce that are probably the same age as the pine, will gradually take their places. Meanwhile balsam fir move into the stand and as black spruce fall out they are replaced by the fir. Fir regenerate prolifically in shade wherever there is a seed source and over the life of practically all stands in our area fir eventually infiltrates. When pine are harvested or die, letting light into the stand, fir will usually be the predominant trees that eventually claim the spaces they leave.

Both trembling aspen (poplar) and white birch will gradually decline as they, too, require near-full sunlight to regenerate and grow. They do not prosper beneath the shade of a partially harvested stand.

Pure black spruce, jack pine, and mixed jack pine and black spruce stands that normally occupy most of NWO's boreal forest will virtually disappear except where wildfires or clear-cut harvesting continue to occur. Some black spruce will regenerate in selection-harvested stands and they will grow, but again fir will more often than not beat them to the best growing spaces and relegate them to a minority position within the stand. I have already mentioned that scraping the forest floor and removing fir seedlings could control fir and enhance black spruce's potential. I rather doubt that such treatments would be economically viable, or to spruce's long-term advantage, as the treatment would have to be repeated at every ten- to twenty-year stand entry. With each treatment we would probably also be eliminating most spruce that had already managed to regenerate.

Prescribed ground fires to control balsam fir ingrowth at thirty- or forty-year intervals may be viable. They would remove all regeneration of all species but would increase seedbed quality for the spruce. Unfortunately the conditions that would make a ground fire possible would create high risk of stand removing crown fires in uneven-aged stands with their inherent ladder fuels. Before such a practice can even be considered, research is needed into the methods, risk of stand loss, and long-term forest health.

White spruce will gradually infiltrate most stands, particularly those on moister, richer soils. They regenerate well in shade, but fir will also relegate them to a minority position in selection-harvested stands. The tiny fir seedlings' coarser roots enable them to utilize a wider variety of micro-sites on the forest floor so that they almost always outnumber white spruce with its finer, less-aggressive roots. Any white spruce that do find the space soon outgrow the fir and, when favoured in selection harvesting, will become dominant, valuable trees.

In summary, the NWO boreal forest managed with selection harvesting will, in my opinion, trend toward a collection of stands, each composed of many-aged primarily balsam fir and white spruce trees.

## CHANGING THE BOREAL FOREST — WE MAY NOT LIKE THE CONSEQUENCES

Let's view what our NWO forest could be like after almost fifty years of wide-scale selection harvesting. I fear that our forest could be disturbingly different from the one that we are familiar with at this time. For that speculative future we assume that our climate will remain similar to that of the past fifty or so years.

In 2065 a selection-harvested forest would probably consist of stands containing predominantly mixed-aged balsam fir and white spruce. In that forest we can anticipate more diseases and insects capitalizing on the new feeding opportunities.

Each time we enter the stands to work, we will damage some of the trees that we are leaving. Even with careful harvesting and extraction we will be bumping and damaging trees, and as larger trees are felled we will be injuring their neighbours. Each injury will be an infection court for disease, and some diseases will become pandemic in the virtual monoculture of balsam fir.

Nature does not allow fir monocultures to exist for long in NWO. When it does, they are usually restricted to relatively small areas, which were either missed by fire or skipped within fire perimeters and later reserved as regulated set-asides from harvest and protected from fire by us.

Long before 2065 our extensive, relatively uniform forest will offer few impediments to the spread of disease, and many trees have a variety of rots long before they mature. Insects, such as spruce budworm, have extensive areas of fir and white spruce to feed on and they are continuing their periodic outbreaks. After each epidemic the stands return to predominantly balsam fir and white spruce. The new stands revert to near-even-aged condition and together cover extensive area. Despite our concerted efforts to convert to multi-aged stands, the natural cycles will tend to return the forest to the even-aged stand structure, but not the species content, that fire and clear-cutting has maintained until this book was published.

Meanwhile the disease- and insect-damaged and killed trees are falling in the face of strong winds. Some blown-down areas extend well into otherwise undamaged stands, and may even be larger than earlier clear-cuts. By 2065 selection harvesting, diseases, insects, and winds combined will have created a forest that is ripe for fire. I fear fires can and, even with a stable climate, will increase in frequency, size, and intensity in such a forest. Even if our climate continues to become damper, the pattern of climate change so far has been erratic and will probably continue to be so. We can't rule out future drought years and, with a preponderance of multi-aged stands and their ladder fuels, even more extreme fires than we have yet experienced.

Among the piles of wind-blown trees and dense fir thickets a few dead and living tall trees remain standing. They are lightning rods and their rotting interiors are vulnerable to ignition. The heavy fuel loads carry resulting fires into otherwise undamaged, multi-aged stands where the flames quickly climb the continuous fuel from ground to treetops. During infrequent droughts large destructive firestorms increasingly plague NWO. Over significant areas, during mid-summer fires, there are now few seed-bearing trees and aspen roots are severely damaged by the abnormally hot fires. Post-fire tree regeneration is limited and what there is consists mostly of birch and aspen. Some areas may even revert to grass or hardwood shrubs.

This scenario will be viewed by some as highly improbable and I will be accused of fear-mongering, but I hope that what I have written will help readers understand my concern.

We need to be aware of the possible consequences of changing the NWO boreal forest away from its time-proven, even-aged (single-aged) stand structure into an uneven-aged-stand (multi-aged) structure that time has shown to be vulnerable to the scenarios that I have described here.

Uneven-aged management and selection harvesting work well in natural, uneven-aged, shade-tolerant hardwood forests of southern Ontario, but to apply the same management system to the natural, even-aged, light-demanding boreal forest of northwestern Ontario is, in my opinion, inviting trouble.

Despite my concern, I recommend that in NWO we begin testing various partial cuts across the range of current stand types and rigorously monitoring natural depletion and stand succession in each. I may be wrong, and if so the results of such trials are needed if we are to assist the boreal forest's transition toward a more southerly forest as our climate warms.

# 7

## FOREST PRACTICES —
## TODAY'S METHODS

Are the things that disturb us about today's timber harvesting more a result of the lingering bad reputation of the tree harvesters rather than the silvicultural techniques being used today? Much of our concern about timber harvesting is based on memories of practices that were common years ago but today are rare. Let's look back four decades.

I recall in the early 1970s participating in national conferences of industrial foresters in which we reflected on rising public criticism of our practices. We knew that media reports and public perceptions were based on a mix of truth and something else, but in the public's eyes, perception was reality. We asked ourselves, "Are we able to make a difference?" The industry was counting on us to propose ways in which it could improve its image. Many of us proposed to our employers that we openly confront our faults and correct them.

Through the 1970s, it remained easy to see careless practices in harvesting operations anywhere in North America. Some of the more obvious practices included:

- ignoring fuel and oil leakage or even intentionally dumping oils on the ground;
- abandoning in the forest obsolete machinery and broken parts;
- leaving broken and retired cable where it was last used;
- excessive bulldozing of rights-of-way and landings on logging roads;
- blocking drainage with poorly constructed roads, culverts, and bridges, and thereby causing soil erosion and stream siltation;
- damaging water quality and fish-spawning beds by using machinery in streams;
- leaving merchantable logs that were lost from skidder loads to rot on cutovers;
- leaving piles of harvested wood to rot at roadsides; and, of course,
- not ensuring regeneration of harvested areas.

This waste of resources and fouling of the natural environment had been ignored and even condoned from the beginning of timber harvesting in Ontario. By the late 1970s, however, it was clear that citizens now cared and they were no longer going to tolerate careless practices. Our faults were exposed for all to see and the industry was being hurt in its markets.

Objective self-examination forced us to admit that there was much we could do to improve. As pressure from lobby groups on our world markets and media attention intensified, our industry began to respond.

To help guarantee future wood supply, the industry requested and successfully negotiated Forest Management Agreements

(FMAs) in which it took responsibility for forest management on a licensed area. With the advent of FMAs during the early 1980s, things began to change. Harvested areas were subjected to independent forest management audits, which opened investors' and management's eyes to the fact that there was much truth in our critics' claims.

My employer empowered its foresters to draft a forest management policy followed by a policy for all of our behaviour in the natural environment. After corporate approval of the final documents, they were made public and applied to company-wide operations. Our executives insisted that all employees from the top down were accountable for living those policies. They communicated that during divisional town-hall meetings and frequent visits to our field operations. I hosted inspections from general managers and vice presidents and felt the pressure of their probing questions and cross-examination. Today, I recall the pride that their commitment generated among us, and how the change in corporate attitude caused dramatic change in the bush. Foresters and all employees were now able to make a difference. We were determined to become leaders for positive change in our industry.

By the mid-1990s, the forest industry had generally cleaned up its act. Oh, yes, there were and still are exceptions to the rule, but a company can no longer expect to ignore public criticism and retain the freedom to carelessly operate without consequences on public land. I am convinced a battered but enlightened industry now welcomes that change.

Now, let's ensure that all other industries, businesses, and citizens follow the forest industry's good example because with objective reflection we can all discover ways to improve our attitudes and practices. Let's do it.

## INDEPENDENT AUDITS — KEY TO GOOD FOREST MANAGEMENT

Are we managing the boreal forest according to the best available current knowledge?

Ontario's Crown-land forests are managed according to regulations and guidelines developed and administered under the *Crown Forest Sustainability Act, 1994*, as most recently amended in 2011. As the introduction to the legislation states, "The purposes of this Act are to provide for the sustainability of Crown forests and, in accordance with that objective, to manage Crown forests to meet social, economic and environmental needs of present and future generations."[1]

The act further provides,

> (3) The Forest Management Planning Manual shall provide for determinations of the sustainability of Crown forests in a manner consistent with the following principles:
>
> 1. Large, healthy, diverse and productive Crown forests and their associated ecological processes and biological diversity should be conserved.
> 2. The long term health and vigour of Crown forests should be provided for by using forest practices that, within the limits of silvicultural requirements, emulate natural disturbances and landscape patterns while minimizing adverse effects on

> plant life, animal life, water, soil, air and
> social and economic values, including rec-
> reational values and heritage values.[2]

When applied to the northwestern Ontario (NWO) boreal forest, those principles effectively require forest managers to try to imitate the patterns and effects of wildfire. There are disturbances in the forest other than fire, but over the long term, when left to nature, fire blurs the patterns of other disturbances. That is to say, in the natural order of events, fire eventually burns the ground on which insects, wind, and disease have depleted stands, and the regenerating species will be those adapted to fire. Nevertheless, those other pre-fire depletions affect seed availability and influence species mixture and stand boundaries in the emerging, post-fire forest.

To satisfy the needs of this legislation and respond to Term and Condition 94 of the original "Class Environmental Assessment (Class EA) for Timber Management on Crown Lands,"[3] the Ministry of Natural Resources and Forestry (MNRF) updated and published detailed silvicultural guides.[4] The guides describe the various forest sites, their associated vegetation groups, plant succession paths, and workable soil and stand treatments. Those guides additionally respond to a number of other policies, including the Environmental Bill of Rights[5] and the MNRF's Statement of Environmental Values.[6]

The guides are based on accumulated science pertaining to managing the boreal forest and the experiences of the range of forest practitioners in our area. Meanwhile, research is continuing on questions brought up in the original Class EA and on new questions that arise as existing guides are implemented.

The guides are thus living documents that are updated with emerging scientific and practical knowledge.

Any commercial tree harvesting on Crown lands in Ontario's boreal forest is practised under MNRF-approved forest management plans, which are developed co-operatively among the licensees' foresters, MNRF's foresters, biologists, and other specialists, and an appointee of the relevant Local Citizens' Committee (LCC). Stand removal and renewal is planned according to the guidelines, which also provide "a framework and a context for generating, collecting, validating and applying local knowledge and experience in the Boreal Forest of Ontario."[7]

I think that when forest management is practised according to Ontario's legislation, regulations, and guides, and tested by independent forest audits according to the MNRF's requirements, then the forest is being managed in the best practical way. Ontario's citizens have the additional assurance that good management is occurring from other audits that forest managers voluntarily submit to from independent international certifying agencies. When deficiencies are revealed by any audit required by either the MNRF or a certifying agency, or both, the licensee is obligated to make specified improvements or suffer the consequences. I am comfortable with that situation, as long as international audits are based on criteria and indicators compatible with our laws and local forest ecosystems and processes. Failure by a certifying agency to accommodate our laws, local environment, and natural forest behaviour could put good forest managers in an untenable position, damage the forest manager's, the certifier's, and their auditor's credibility, and, most important, the health of the forest.

## ACKNOWLEDGING THE FOREST INDUSTRY'S
## IMPROVEMENT AND HARNESSING ITS CAPABILITIES

Over the course of my career, I was employed by a series of Canadian forest industry companies that had forward-thinking owners and managers. Some were leaders in responsible forest management for their time and ahead of legislated requirements. Some senior industrial executives and managers I have known held themselves to high standards based on their ideals. Yes, some believed their companies had roles to play in moving Canada and the world toward more responsible forest and environmental management.

The world is very different today. Long-term demand for newsprint has declined in response to advances in and use of electronic communication. Meanwhile, the worldwide economic slump depressed demand for other forest products. It has been difficult for forest product companies to adjust to the new reality. However, despite the new situation they find themselves operating in, Canadian and Ontario forest product companies continue to use scarce cash to further reduce their environmental footprints and greenhouse gas emissions. Their extraordinary record to date for reducing $CO_2$ release from their operations makes them unique among Canadian industries and, yes, governments, too.

I am afraid that environmental activists who are targeting markets of progressive companies are alienating those companies from trying to improve further. If I am already one of the best and I am one of the ones being harassed, the leaders of such companies may say to themselves, what gain is there for me to improve further? Other companies and their managers will see the leaders of these progressive companies being penalized and will be even more discouraged and cynical. If so, we have a serious problem. There has to be a better way to encourage investment in a greener economy.

Over the past fifty years, the Canadian forest industry has responded well to lobbies against demonstrably poor practices, and it continues to do so. The time has come for precise, constructive action. We should all self-examine and, as the forest industry is doing, when we discover deficiencies in our behaviours, buckle down and make the necessary changes. We can all follow their example and adopt more positive and co-operative behaviours.

Us-versus-them confrontations are now wasting precious time and deepening our mess. More of the same is not going to save us.

We need to look for the positives in others' capabilities and assertively co-operate with them. Forest management is probably the best place to start because there is already a collective will to continue improving. We have good legislation and agreements that are working. Funding of forest renewal after harvesting and disaster is guaranteed by Forest Renewal and Forestry Futures Trust Funds that are financed by industry under government regulation. Credible independent agencies are auditing forest management and there is an improving track record. We should not ignore poor practices; instead, they should be highlighted and the full weight of law applied to proven offenders. Good legislation and the force of law, not vigilantism, should be the tools that force bad actors into line.

Like it or not, in our democracy governments don't have the money, will, or long-term tenure to consistently move on environmental issues. Private investors do, however, have the ability to apply long-term solutions — when clear and consistent policy is applied. They also have more cash than governments have and they can quickly apply it to a targeted need.

The forest industry has in the past had, and so far it still has, committed leaders willing and able to move quickly and effectively, given the necessary political and public support.

We can all do better. For the sake of the Earth and our descendants, let's all start working together in our common interest.

# CONCLUSION

I enjoyed both the writing of and my reader's reactions to the articles that were published in the *Chronicle Journal*. The articles described the boreal forest as I understand it, based on my training and life experiences. More recently, I have enjoyed writing *Dynamic Forest*. It has been a new experience, one that has added challenge and excitement in my declining years.

Over the past three decades, I have had a growing concern that Earth is nearing the end of its tolerance for our excessive abuse of our life-support systems. My readings of learned articles on the subject, and witnessing weather event and ecosystem changes have convinced me that climate change is upon us and that we are the primary cause. Some say that the way we are managing the boreal forest is a major cause of ecosystem and climate change. I disagree. I fear that if we change forest management based on popular perceptions we may only deepen the crisis.

With the helpful understanding and tolerance of some wise individuals, I was given the scope to research my biases and question my objectivity. With their support, I studied, and wrote or supervised concluding reports on some of the oldest consistently maintained harvesting, renewal, and stand-management trials in the Canadian boreal forest. Later observation of several stands in NWO over the next three decades helped confirm for me that my earlier findings also apply there. That experience has shown me that if we are to sustain and continue living and working within the boreal forest we must control forest fires as best we can, and harvest and renew the forest to resemble the landscape pattern, species content, and even-aged structure that follows wildfire.

Experience has been a powerful influence on my appreciation of and attitude toward utilization of the boreal forest. I changed from a youth who was disgusted with clear-cutting into a forester who prescribed clear-cutting because I learned that when done well it works. My experience has led me to suspect that it works better than other harvesting alternatives, but we still need more-rigorous scientific trials of those alternatives to prove that.

Some critics of Ontario's management of the boreal forest oppose extensive single-species plantations as dangerous monocultures that invite wide-scale disease and insect epidemics. However, foresters know that in NWO's fire-driven boreal forest monocultures are common and part of the natural biodiversity. In fact, natural mixed stands of balsam fir and white spruce with some black spruce, the so-called boreal old-growth condition, are more likely to experience depletion by insects. On the other hand, we have found other tree species infiltrating so-called monoculture plantations. In time, the plantations tend to become more like the commoner, predominantly coniferous, stands of jack pine and black spruce with some trembling aspen and white birch that follow most wildfires. Stands with mixtures of those species, whether they are natural or artificial, are relatively resistant to insect epidemics.

Well-meaning individuals and organizations continue to criticize boreal foresters for prescribing clear-cut harvesting instead of selection harvesting. I caution against creating a preponderance of uneven-aged selection harvested stands because they will contain few, if any, trees capable of regeneration following most wildfires. I fear that a forest that is composed mainly of man-made uneven-aged stands in NWO will be extremely vulnerable to large uncontrollable firestorms like never before witnessed. That forest will be unfit to regenerate to trees after most fires, and the fires will cause unprecedented loss of life and property.

Horse logging has been advocated by some, but the light soil disturbance of such logging allows the balsam fir and white spruce advanced growth to be retained in the cutover whether it be clear-cut or partial-cut. Many of the stands that were killed in the spruce budworm epidemic of the 1980s developed on such earlier horse logged cutovers.

Machine logging, particularly in clear-cutting, crushes and kills much of the fir advanced growth and reduces the probability of future stand vulnerability to spruce budworm epidemics. Spruce advanced growth in NWO is seldom plentiful enough, except on lower-quality sites such as rocky areas, to contribute much to the next stand. Careful harvesting that protects the spruce, where numbers warrant, may be worth the effort. Those soils usually support slower growth and consequent lower yield, thus increased harvesting cost may be a better return on investment than is artificial regeneration.

Ontario legislation and regulations have effectively enabled the development of forest management planning manuals and guides around forest management. Trust funds that receive payments from timber licensees and fund forest renewal and some other aspects of management help ensure forest renewal and protection. Adherence to plans is supported by independent forest management, audits, and penalties for failure to correct deficiencies. Dramatic improvement in licensees' behaviour over the past four decades has helped place Ontario's forests among the best managed in the world. However, more improvement can be made.

Ontario has decided that its future forests should be sustainable and resemble the forests that follow natural disturbance. Fire has historically created the forest stand pattern, structure, and species content in NWO, and that is the model that modern forest managers are attempting to imitate. Foresters know that we can never imitate exactly natural processes, but by learning from our mistakes we can continue to improve. I am convinced that improvement will come easier and faster if all stakeholders and concerned parties can find common ground and co-operate.

Ken Armson, a man who has arguably had more positive impact on Ontario forestry and forests than any other individual, recently said, "… a major activity in forestry will always be some form of cutting trees whether or not for timber or other values the focus from the outside is still on 'logging' and not on the larger picture of forest management. The challenge is to develop a public perspective of the management of Canada's forests, i.e., forestry, which in time will become the essence of our culture and forest history."[1]

I hope *Dynamic Forest* helps advance that public perspective.

# ACKNOWLEDGEMENTS

Individuals who helped me during the series of articles in the *Thunder Bay Chronicle Journal* and subsequently the preparation of this book include Kip Miller of the Ministry of Natural Resources and Forestry (MNRF), Thunder Bay District Fire Centre, Thunder Bay Fire Chief John Hay, and City Forester Shelley Vescio, who collectively encouraged me to begin the series by focusing on the FireSmart program for the city. Jolene Davis, a former columnist with the *Chronicle Journal*, advised me on a format for my approach to the editors and encouraged me on the basis of her familiarity with my drawings and writing. Julio Gomes, the former managing editor of the *Chronicle Journal*, and Greg Giddens, the current one, gave me the forum and guidance on style. Werner Schwar, a local landscape architect, advised me early in the series on my choice of words to ensure I got my intended message across without causing overreaction. Kevin Ride of the MNRF helped update me on current forest policy.

I am grateful to five former workmates in particular at Abitibi-Price Inc. in Thunder Bay. Each helped familiarize me with the local boreal forest and the subtleties of its effective management. Brian Cavanaugh, Volker Kromm, Paul Poschmann, Bill Smith, and, for a short period, Jack Winkler were the real forest managers; I was a hanger-on.

A good friend, Peter Nicholas, a local land manager and former working partner at Abitibi-Price, helped during the writing of the newspaper articles with his knowledge of current local forest operations, gave constant encouragement and feedback on reader reactions, offered suggestions on future topics, and helped with editing an early draft of the manuscript.

Muriel Squires, the best English teacher a husband ever had, corrected my punctuation, spelling, and grammar in the original articles, and put up with the strange and endless hours I devoted to this project. Many lay and professional readers, too numerous to mention, gave critical informed opinions through email, telephone exchanges, conversations at public gatherings, in my barber's chair, at store checkouts, and on the street. Knowing that I was connecting with a diverse readership was encouraging. Receiving their views, questions, and suggestions helped me to add substance and clarity to subsequent articles. For three decades I have benefited from the friendship, insight, practical and expert advice, constructive criticism, and encouragement of Dr. Mark Kuhlberg, professional tree planter, knowledge and truth sleuth, professor of history at Laurentian University, and chair of the Forest History Society of Ontario. Most recently, Mark took time out from his intense schedule to edit the draft manuscript before it was submitted to Dundurn.

Of course, you would not be reading *Dynamic Forest* if it were not for J. Patrick Boyer, Editor-at-Large, Politics and History and General Editor, Point of View books at Dundurn Press. He showed immediate interest based on a sketchy description of an early draft manuscript. Through his patience and early guidance, what was once a series of disconnected articles has morphed into this book. The theme has remained constant, the content has become more focused, and the structure has changed. I am grateful to Patrick and Kirk Howard, Publisher at Dundurn Press, for their confidence in me and the project, and to the team at Dundurn Press for helping me through one of the most gratifying experiences of my long life.

All of the following individuals at Dundurn gave me invaluable advice and encouragement as I struggled to meet their high standards and produced a book that they must feel is worthy of their support:

- Laura Boyle, Senior Designer;
- Margaret Bryant, Director of Sales and Marketing;
- Sheila Douglas, Executive Assistant and Manager; Contracts and Administrative Services;
- Dominic Farrell, Developmental Editor;
- Carrie Gleason, Editorial Director;
- Kyle Gray, Marketing Coordinator;
- Kathryn Lane, Managing Editor;
- Synora Van Drine, Sales Manager;
- Cheryl Hawley, Production Editor;
- Catherine Leek, freelance copy editor; and
- Kendra Martin, Publicist and Marketing Administrator.

Dominic Farrell's editing guided me through numerous changes to the manuscript, not only the smaller things, such as wording and punctuation, but also some larger issues, such as additions and revisions of the structure — all necessary to improve your understanding of what I am trying to communicate to you. He helped me change my writing from what was once described by one of my professors as "an example of how not to write an essay" into something that eventually passed muster with Dundurn.

Cheryl Hawley had the difficult task of prodding me to adhere to the production schedule despite my tendency to take longer than should be expected for even the easy jobs.

Catherine Leek, my copy editor, put the final brush-up on my punctuation, spelling, word choice, and inconsistent note format.

I feel that all have become my friends. I reserve the last appreciation for special recognition of Dr. John Naysmith R.P.F. (Ret.), founding dean of the Faculty of Forestry (now the Faculty of Natural Resources Management) at Lakehead University, for his unwavering encouragement and support throughout the series and most recently for his generous guidance and help in getting this book prepared and published.

Thank you all for your generous support and encouragement.

# NOTES

## INTRODUCTION

1. During the summer of 1959 I worked with a forestry survey crew evaluating the forest of the newly created Terra Nova National Park. At that time it was already evident that moose were heavily browsing balsam fir and hardwood species' regeneration on recently disturbed locations. I have since visited the park several times and witnessed a conversion of some of those once-productive forest sites to virtual grassland, bracken fern thicket, or near-kalmia heath.

## CHAPTER 1: CANADA IS A FOREST NATION

1. Natural Resources Canada, *The State of Canada's Forests: Annual Report 2015* (Ottawa: NRC, 2015), 51.
2. Food and Agriculture Organization of the United Nations, *Global Forest Resources Assessment 2015: How Are the World's*

*Forests Changing?*, 2nd ed. (Rome: FAO, 2015), 3, www.fao.org/3/a-i4793e.pdf.

3. Natural Resources Canada, *The State of Canada's Forests*, 50.

4. Ibid.

5. Hannes Mäntyranta, "Clearcut Areas Are Astonishingly Small," Finnish Forest Association, last modified November 4, 2014, www.smy.fi/en/artikkeli/clearcut-areas-are-astonishingly-small.

6. K. Kolis, J. Hiironen, E. Ärölä, and A. Vitikainen, "Effects of Sale-Specific Factors on Stumpage Prices in Finland," *Silva Fennica* 48, no. 3 (2014): 2.

7. "Forest Certification in Canada," National Resources Canada, last modified November 22, 2016, www.nrcan.gc.ca/forests/canada/certification/17474.

8. "Facts About Wildland Fires in Canada," Natural Resources Canada, last modified September 6, 2016, www.nrcan.gc.ca/forests/fire-insects-disturbances/fire/13143.

9. "Forestry Highlights, 2015," National Forestry Database, http://nfdp.ccfm.org/highlights/highlights_e.php; "Silviculture, 2015," National Forestry Database, http://nfdp.ccfm.org/silviculture/quick_facts_e.

10. Lars Rytter, Morten Ingerslev, Antti Kilpeläinen, Piritta Torssonen, Dagnija Lazdina, Magnus Löf, Palle Madsen, Peeter Muiste, and Lars-Göran Stener, "Increased Forest Biomass Production in the Nordic and Baltic Countries — A Review on Current and Future Opportunities," *Silva Fennica* 50, no. 5 (2016): 3–4.

11. "Grassy Narrows Hosts Gathering to Highlight Concerns Over Logging," *Chronicle Journal*, July 11, 2006.

12. Bryan Meadows, "Clearcuts Under Fire Again," *Chronicle Journal*, March 1, 2005.

13. Juli G. Pausas and Jon E. Keeley, "A Burning Story: The Role of Fire in the History of Life," *BioScience* 59, no. 7 (2009): 593–601.

## CHAPTER 2: HOW I BECAME A FORESTER

1. I recommend a book by John Kitchen, *By the Sweat of My Brow: The Life of a Newfoundland Logger* (St. John's, NL: John Kitchen, 2005), 143–58. In my opinion, Kitchen gives an accurate review of events leading up to and during the International Woodworkers of America (IWA) strike, during which deep emotions often resulted in actions individuals would not otherwise have taken. Indeed, during one incident a member of the Royal Newfoundland Constabulary was killed. As a university student in 1958, I attended a session of the Newfoundland Provincial Parliament when Premier Smallwood delivered a long monologue, to an almost empty house, about the IWA and its "unlawful actions." He made frequent reference to all the loggers in the visitor gallery, but I and my two student friends were the only persons that I recall being there. He ultimately prevailed and the government decertified the IWA in the province.

## CHAPTER 3: THE BOREAL FOREST NEEDS SOUND SCIENCE

1. These details are from my memory of an internal report that I wrote at Price (Nfld.) Pulp & Paper Limited, describing

the design, execution, analysis, and conclusions of a regeneration survey of the Noel Paul River watershed in Central Newfoundland. The area had been clear-cut harvested over a period from the 1930s to the early 1950s and closed to the public by virtue of a company-owned railway, which provided the only transportation to and from its divisional logging locations. The report details the successful regeneration of the species harvested, but declares some areas of the watershed to have later become devoid of tree growth except for scattered young white spruces and residual old white birches.

During spring and summer moose were ranging widely outside the area, but during the fall hunting season the animals were moving into the inaccessible Noel Paul River watershed, presumably to avoid hunters. The result was a winter population of approximately twenty moose per square mile (seven or eight per square kilometre). The regeneration survey concluded that on a significant portion of the watershed, moose were eliminating balsam fir, white birch, trembling aspen, mountain ash, red and mountain maple, red-osier dogwood, other hardwood shrubs, and ground hemlock (Canada yew), and the area was converting to virtual grassland.

2.  From a 1969 internal final report on the termination of a pre-commercial thinning trial. In 1921 four blocks had been laid out in an approximately twenty-year-old burn on a flood plain of the Exploits River, Newfoundland. One block was retained as a control and the other three blocks were thinned to three different spacings among mostly black spruce saplings. The report, based on forty-eight years of growth after the thinning, concluded that the

six-by-six-foot (1.83-by-1.83-metre) spacing block showed significantly superior growth in merchantable volume to that of the other three blocks and superior tree size to that of the control. The two wider-spaced plots experienced significant recovery of supposedly killed spruces and ingrowth of firs that competed with the reserved trees, restricting their growth. The effect on those two blocks was that stand volume growth was similar to that of the control.

3. Philip Mathias, *Takeover: The 22 Days of Risk and Decision That Created the World's Largest Newsprint Empire, Abitibi-Price* (Toronto: Maclean-Hunter, 1976). Having, for several years, known most of the key players on the Price Brothers Limited side of this story, and after twice meeting and having discussions with the CEO and the president of Abitibi, the book had special appeal to me. It filled in details that connected some events with which I was familiar and gave me some limited insight into high finance and boardroom discussions.

4. Geomatics International, *An Independent Forest Audit of the White River Forest SFL 501500 for the Period 1993 to 1998* (prepared for Ministry of Natural Resources, Corporate Affairs Branch, Audit and Evaluation Section, 1999). This is one of two independent forest-management audits for which I was the lead auditor.

5. AWH Damman, "Some Forest Types of Central Newfoundland and Their Relationship to Environmental Factors," *Forest Science*, no. 8 (1964): 62.

6. Mac Squires, "Observations of a Plantation — The Anders Plantation in the Black Spruce Forest," *Forestory* 3, no. 1 (2012): 66.

7.  *Canadian Boreal Forest Agreement — Abridged Version*, http://cbfa-efbc.ca/wp-content/uploads/2014/12/CBFA Agreement_Abridged_NewLook.pdf.
8.  See "Canadian Boreal Forest Agreement," *Nishnawbe Aski Nation*, www.nan.on.ca/article/canadian-boreal-forest-agreement-462.asp.
9.  Professional Foresters Association Act, 1957, S.O. 1957, c. 149.
10. Professional Foresters Act, 2000, S.O. 2000, c. 18.
11. Ibid., O. Reg. 145/01.
12. W.D. Bakowsky, *Prairies and Savannahs of Northwestern Ontario* (Peterborough, ON: Natural Heritage Information Centre, 2007), 3.
13. Crown Forest Sustainability Act, R.S.O. 1990, c. 25.
14. Ministry of Natural Resources, *Silvicultural Guide to Managing for Black Spruce, Jack Pine and Aspen on Boreal Forest Ecosites in Ontario*, ver. 1.1 (Toronto: Queen's Printer for Ontario, 1997).
15. Ministry of the Environment and Energy, *Decisions and Reasons for Decision, Class Environmental Assessment by the Ministry of Natural Resources for Timber Management on Crown Lands in Ontario* (Toronto: Environmental Assessment Board, 1994), 561.
16. Personal communication from David Jones.
17. Personal communication from Maurice Rubenick.

## CHAPTER 4: EACH SPECIES HAS UNIQUE REQUIREMENTS

1.  Mac Squires, "Observations of a Plantation — The Anders Plantation in the Black Spruce Forest," *Forestory* 3, no. 1 (2012): 33.

2. Larry Watkins, *The Forest Resources of Ontario 2011.* (Toronto: Queen's Printer for Ontario, 2011), 5.

3. John Laird Farrar, *Trees in Canada* (Markham, ON: Fitzhenry and Whiteside, 1995), 107.

4. Squires, "Observations of a Plantation," 34.

5. V.G. Smith, M. Watts, and D.F. James, "Mechanical Stability of Black Spruce in the Clay Belt Region of Northern Ontario," *Canadian Journal of Forest Research* 17, no. 9 (1987): 1088.

6. Farrar, *Trees in Canada*, 284–86.

7. Ibid., 286.

8. The North House Folk School in Grand Marais, Minnesota, offers numerous courses in folk crafts and art from birch bark canoe construction, cooperage, and bread making to bird identification and mushroom collection. Having taken a course there and being familiar with them for a number of years, I highly recommend them to anyone wanting to reconnect with the necessities and skills of pioneer life. Check them out at https://northhouse.org.

9. Mac Squires, "The Dynamic Forest — Firewood Users Deserve Spin-Free Answers," *Chronicle Journal*, October 18, 2014.

10. Farrar, *Trees in Canada*, 347.

11. Ibid., 102.

12. Ibid., 26.

13. Ibid., 74.

14. Ibid., 44.

15. Ibid.

16. Ibid., 45.

17. Ibid., 166.

18. Ibid., 56.
19. Northwestern Ontario Forest Technology Development Unit, *Forest Research and Management Demonstration Area: Thunder Bay Spacing Trial* (Toronto: Queen's Printer for Ontario, 1989), 10.

## CHAPTER 6: HARVESTING THE BOREAL FOREST

1. Mäntyranta, "Clearcut Areas Are Astonishingly Small."
2. Ibid.
3. Ministry of Natural Resources, *Silvicultural Guide*, 1:14.
4. Ibid., 1:19.
5. Ibid.
6. Raveena Aulakh, "Ontario Gives Green Light to Clear-Cutting at Grassy Narrows," *Toronto Star*, December 29, 2014.

## CHAPTER 7: FOREST PRACTICES — TODAY'S METHODS

1. Crown Forest Sustainability Act, 1994, S.O. 1994, c. 25, s. 1.
2. Ibid., s. 2(3).
3. Ministry of the Environment and Energy, *Decisions and Reasons for Decision*, 561.
4. Ministry of Natural Resources, *Silvicultural Guide*.
5. Environmental Bill of Rights, 1993, S.O. 1993, c. 28.
6. "Statement of Environmental Values: Ministry of the Environment and Climate Change," Environmental Registry, www.ebr.gov.on.ca/ERS-WEB-External/content/sev.jsp?pageName=sevList&subPageName=10001.
7. Ministry of Natural Resources, *Silvicultural Guide*, 1:viii.

## CONCLUSION

1. K.A. Armson, "Canadians and Their Forests," *Working Forest* 20, no. 8 (2016), 21.

# BIBLIOGRAPHY

The following list of references covers the range of topics addressed in this book. Some delve into a part of the forest or an event and add to our understanding of the behaviour of the forest. Others combine the findings of several such studies on a variety of subjects into practical field guides for foresters.

Ontario has an active forest research program on a variety of subjects. The province makes excellent use of its findings and those of others in developing its useful forest management manual and related guides, some of which are included in the following list.

Allison, Tammy, and Henry Michel. "Helping Our Land Heal: A Cultural Perspective on Fire and Forest Restoration." *BC Grasslands* (Fall 2004): 7.

Bakowsky, W.D. *Prairies and Savannahs of Northwestern Ontario.* Peterborough, ON: Natural Heritage Information Centre, 2007.

Baskerville, G.L. "Spruce Budworm: The Answer Is Forest Management: Or Is It?" *Forestry Chronicle* 51, no. 4 (1975).

Bell, F. Wayne. *Critical Silvics of Conifer Crop Species and Selected Competitive Vegetation in Northwestern Ontario*. Thunder Bay, ON: Ministry of Natural Resources, 1991.

Bentley, Cathy, and Fred Pinto. *The Autecology of Selected Understory Vegetation in Central Ontario*. North Bay, ON: Ministry of Natural Resources, 1994.

Bergeron, Yves, Mike Flannigan, Sylvie Gauthier, Alain Leduc, and Patrick Lefort. "Past, Current and Future Fire Frequency in the Canadian Boreal Forest: Implications for Sustainable Forest Management." *Ambio* 33, no. 6 (2004): 356–60.

Bergqvist, Göran, Roger Bergström, and Martha Wallgren. "Recent Browsing Damage by Moose on Scots Pine, Birch and Aspen in Young Commercial Forests — Effects of Forage Availability, Moose Population Density and Site Productivity." *Silva Fennica* 48, no. 1 (2014).

Bowling, C., and G. Niznowski. *Factors Affecting Jack Pine Cone and Seed Supply After Harvesting in Northwestern Ontario*. Thunder Bay, ON: Ministry of Natural Resources, 1991.

Bowling, C., G. Niznowski, and M. Maley. *Ingress of Natural Regeneration in Plantations after Tree-Length Harvest in Northwestern Ontario*. Thunder Bay, ON: Ministry of Natural Resource, 1997.

Brassard, Brian W., and Han Y.H. Chen. "Stand Structural Dynamics of North American Boreal Forests." *Critical Reviews in Plant Sciences* 25, no. 2 (2006): 115–37.

Burton, P.J. et al, eds. "Sustainability of Boreal Forests and Forestry in a Changing Environment." *IUFRO World Series* 25 (2010): 249–82.

Buse, L.J., and W.D. Baker. *Black Spruce Site Quality for the Northwestern Region of Ontario*. Thunder Bay, ON: Ministry of Natural Resources, 1991.

Buse, L.J., and J. Farnsworth. *Black Spruce Advance Growth: Its Potential in Northwestern Ontario*. Thunder Bay, ON: Ministry of Natural Resources, 1995.

Carmean, W.H. *Site Quality Evaluation, Site Quality Maintenance, and Site-Specific Management for Forest Land in Northwest Ontario*. Thunder Bay, ON: Ministry of Natural Resources, 1997.

Caveney, E.W., and Victor J. Rudolph. *Reproducing Jack Pine by the Shelterwood Method*. East Lansing, MI: Michigan State University, 1970.

Chrosciewicz, Z. "Jack Pine Regeneration After Scattering Slash on Exposed Mineral Soil." *Pulp and Paper* 61, no. 3 (1960): 3–4.

Daniel, T.W., J.A. Helms, and F.S. Baker. *Principles of Silviculture*. 2nd ed. Toronto: McGraw-Hill, 1979.

Dawkins, Richard. *The Greatest Show on Earth: The Evidence for Evolution*. New York: Free Press, 2009.

DeWoody, Jennifer, Carol A. Rowe, Valerie D. Hipkins, Karen E. Mock. "'Pando' Lives: Molecular Genetic Evidence of a Giant Aspen Clone in Central Utah." *Western North American Naturalist* 68 (2008): 493–97.

Dohlberg, Anders. "Effects of Fire on Ectomycorrhizal Fungi in Fennoscandian Boreal Forests," *Silva Fennica* 36, no. 1 (2002), 69–80.

Elie, Jean-Gabriel, and Jean-Claude Ruel. "Windthrow Hazard Modelling in Boreal Forests of Black Spruce and Jack Pine." *Canadian Journal of Forest Research* 35, no. 11 (2005): 2655–63.

Elling, Arthur E., and Elon S. Verry. "Predicting Wind-Caused Mortality in Strip Cut Stands of Peatland Black Spruce." *Forestry Chronicle* 54, no. 5 (1978): 249–52.

Fabra Crespo, Miguel, Olli Saastamoinen, Jukka Matero, and Hannes Mäntyranta. "Perceptions and Realities: Public Opinion on Forests and Forestry in Finland, 1993–2012." *Silva Fennica* 48, no. 5 (2014).

Farrar, John Laird. *Trees in Canada.* Markham, ON: Fitzhenry and Whiteside, 1995.

Fenton, Nicole J., Martin Simard, and Yves Bergeron. "Emulating Natural Disturbances: The Role of Silviculture in Creating Even-Aged and Complex Structures in the Black Spruce Boreal Forest of Eastern North America." *Journal of Forest Research* 14, no. 5 (2009): 258–67.

Fleming, R.L., and R.M. Crossfield. *Strip Cutting in Shallow-Soil Upland Black Spruce Near Nipigon, Ontario. III. Windfall and Mortality in the Leave Strips, Preliminary Results.* Sault Ste. Marie, ON: Canadian Forestry Service, 1983.

Fleming, R.L., and D.S. Mossa. "Seed Release from Black Spruce Cones in Logging Slash." *Canadian Journal of Forest Research* 26, no. 2 (1996): 266–76.

Fuglem, G., M. Sabourin, and S. Lundqvist. "Influence of Spruce Wood Properties on Thermomechanical Pulping-Pilot Scale Results." *Proceedings of International Mechanical Pulping Conference*, Quebec City, May 2, 2003.

Geomatics International. "An Independent Forest Audit of the White River Forest SFL 501500 for the Period 1993 to 1998," For Ontario Ministry of Natural Resources, Corporate Affairs Branch, Audit and Evaluation Section, 1999.

Goble, B.C., and C. Bowling. *Five-year Growth Response of Thinned Jack Pine Near Atikokan, Ontario.* Thunder Bay, ON: Ministry of Natural Resources, 1993.

Greene, D.F., J.C. Zasada, L. Sirois, D. Kneeshaw, H. Morin, I.

Charron, and M.-J. Simard. "A Review of the Regeneration Dynamics of North American Boreal Forest Tree Species." *Canadian Journal of Forest Research* 29, no. 6 (1999): 824–39.

Groot, Arthur. "Is Uneven-Aged Silviculture Applicable to Peatland Black Spruce (Picea mariana) in Ontario, Canada?" *Forestry* 75, no. 4 (2002): 437–42.

———. *Silvicultural Consequences of Forest Harvesting on Peatlands: Site Damage and Slash Conditions.* Sault Ste. Marie, ON: Canadian Forestry Service, 1987.

———. "Silviculture Systems for Black Spruce Ecosystems." In *Proceedings: Innovative Silvicultural Systems in Boreal Forests,* edited by Colin R. Bamsey, 47–51. Edmonton, AB: Clear Lake, 1994.

Groot, Arthur, and Derek W. Carlson. "Influence of Shelter on Night Temperatures, Frost Damage, and Bud Break of White Spruce Seedlings." *Canadian Journal of Forest Research* 26, no. 9 (1996): 1531–38.

Groot, Arthur, Derek W. Carlson, Robert L. Fleming, and James E. Wood. *Small Openings in Trembling Aspen Forest: Microclimate and Regeneration of White Spruce and Trembling Aspen.* Sault Ste. Marie, ON: Canadian Forestry Service, 1997.

Groot, Arthur, and C.R. Mattice. *Second Growth Black Spruce Stands on Peatlands Provide Lessons for Current Silviculture.* Sault Ste. Marie, ON: Canadian Forestry Service, 1995.

Harrison, R. Bruce, Fiona K.A. Schmiegelow, and Robin Naidoo. "Stand-Level Response of Breeding Forest Songbirds to Multiple Levels of Partial-Cut Harvest in Four Boreal Forest Types." *Canadian Journal of Forest Research* 35, no. 7 (2005): 1553–67.

Hobson, K.A., and Jim Schieck. "Changes in Bird Communities in Boreal Mixedwood Forest: Harvest and Wildfire Effects over 30 Years." *Ecological Applications* 9, no. 3 (1999): 849–63.

Homagain, Krish, Chander Shahi, Willard Carmean, Mathew Leitch, and Colin Bowling. "Growth and Yield Comparisons for Red Pine, White Spruce and Black Spruce Plantations in Northwestern Ontario." *Forestry Chronicle* 87, no. 4 (2011).

Horton, K.W., and E.J. Hopkins. *Influence of Fire on Aspen Suckering*. Canadian Department of Forestry, Forest Research Branch, publication no. 1095 (1964).

Jeglum, J.K. *Strip Cutting in Shallow-Soil Upland Black Spruce Near Nipigon, Ontario. IV. Seedling-Seedbed Relationships*. Sault Ste. Marie, ON: Department of Forestry, 1984.

Jeglum, J.K., and D.J. Kennington. *Strip Clearcutting in Black Spruce: A Guide for the Practicing Forester*. Sault Ste. Marie, ON: Forestry Canada, 1993.

Joyce, D. *Seed Zones for Ontario [GIS produced map]*. Sault Ste. Marie, ON: Ministry of Natural Resources, 1995.

Kemperman, G.A. *Aspen Clones: Development, Variability and Identification*. Ontario: Ministry of Natural Resources, 1977.

Kitchen, John. *By the Sweat of My Brow: The Life of a Newfoundland Logger*. St. John's, NL: John Kitchen, 2005.

Kneeshaw, Daniel, and Sylvie Gauthier. "Old Growth in the Boreal Forest: A Dynamic Perspective at the Stand and Landscape Level." *Environmental Reviews* 11, no. S1 (2003).

Kneeshaw, D.D., B.D. Harvey, G.P. Reyes, M.N. Caron, and S. Barlow. "Spruce Budworm, Windthrow and Partial Cutting: Do Different Partial Disturbances Produce Different Forest Structures?" *Forest Ecology and Management* 262, no. 3 (2011): 482–90.

Kulhberg, Mark. *In the Power of the Government, the Rise and Fall of Newsprint in Ontario, 1894-1932*. Toronto: University of Toronto Press, 2015.

Kuuluvainen, T., and Grenfell, R. "Natural Disturbance Emula-
tion in Boreal Forest Ecosystem Management — Theories,
Strategies, and a Comparison with Conventional Even-Aged
Management." *Canadian Journal of Forest Research* 42, no. 7
(2012): 1185–203.

Lancaster, Kenneth F. "White Pine Management: A Quick Review." Durham, NH: USDA Forestry Services, 1984.

Losee, STB. "Results of Group Cutting for Black Spruce Regeneration at the Abitibi Woodlands Laboratory." *Pulp Paper* (1961).

———. "Strip Group Cutting in Black Spruce at the Abitibi
Woodlands Laboratory." *Pulp Paper* (June 1966).

Lundmark, Tomas, and Jan-Erik Hällgren. "Effects of Frost on
Shaded and Exposed Spruce and Pine Seedlings Planted in the
Field." *Canadian Journal of Forest Research* 17, no. 10 (1987):
1197–201.

Mathias, Philip. *Takeover: The 22 Days of Risk and Decision That
Created the World's Largest Newsprint Empire, Abitibi-Price.*
Toronto: Maclean-Hunter, 1976.

McCarthy, John. "Gap Dynamics of Forest Trees: A Review with
Particular Attention to Boreal Forests." *Environmental Reviews*
9, no. 1 (2001): 1–59.

Messier, Christian, René Doucet, Jean-Claude Ruel, Yves Claveau,
Colin Kelly, and Martin J. Lechowicz. "Functional Ecology of
Advance Regeneration in Relation to Light in Boreal Forests."
*Canadian Journal of Forest Research* 29, no. 6 (1999): 812–23.

Methven, Ian R., and W.G. Murray. "Using Fire to Eliminate
Understory Balsam Fir in Pine Management." *Forestry
Chronicles* 50, no. 2 (1974): 77–79.

Ministry of the Environment and Energy. *Decisions and Reasons
for Decision, Class Environmental Assessment by the Ministry of*

*Natural Resources for Timber Management on Crown Lands in Ontario*. Toronto: Environmental Assessment Board, 1994.

Ministry of Natural Resources. *Forest Management Planning Manual for Ontario's Crown Forests*. Toronto: Queen's Printer for Ontario, 1996.

———. *Jack Pine Working Group: Silvicultural Guide Series*. Toronto: Queen's Printer for Ontario, 1994.

———. *Spruce Budworm Management Strategy for Ontario*. Toronto: Queen's Printer for Ontario, 1988.

Morris, D.M., C. Bowling, and S.C. Hills. "Growth and Form Responses to Precommercial Thinning Regimes in Aerially Seeded Jack Pine Stands: 5th Year Results." *Forestry Chronicles* 70, no. 6 (1994): 780–87.

Nevalainen S., J. Matala, K.T. Korhonen, A. Ihalainen, and A. Nikula. "Moose Damage in National Forest Inventories (1986–2008) in Finland." *Silva Fennica* 50, no. 2 (2016).

Oliver, Chadwick D., and Bruce C. Larson. *Forest Stand Dynamics*. New York: McGraw-Hill, 1990.

Ontario Forest Policy Panel. *Diversity: Forests, People, Communities; A Comprehensive Forest Policy Framework for Ontario*. Toronto: Queen's Printer for Ontario, 1993.

Place, ICM. "The Influence of Seed-Bed Conditions on the Regeneration of Spruce and Balsam Fir." *Department of Northern Affairs and National Resources*. Forestry Branch Bulletin no. 117 (1955).

Power, Randal G. *Vegetation Management Plan for Terra Nova National Park 2000–2004*. Heritage Integrity Terra Nova National Park, Parks Canada, 2000.

Riley, L.F. *The Effect of Seeding Rate and Seedbed Availability on Jack Pine Stocking and Density in Northeastern Ontario*. Sault Ste. Marie, ON: Canadian Forestry Service, 1980).

Robinson, A.J. *Spruce Regeneration Resulting from Seed Tree Cutting and Clearcutting in Newfoundland.* St. John's, NL: Robinson Newfoundland Forest Research Centre, 1970.

Ruel, Jean-Claude, Christian Messier, Yves Claveau, René Doucet, and Phil Comeau. "Morphological Indicators of Growth Response of Coniferous Advance Regeneration to Overstory Removal in the Boreal Forest." *Forestry Chronicle* 76, no. 4 (2000): 633–42.

Rytter, L., M. Ingerslev, A. Kilpeläinen, P. Torssonen, D. Lazdina, M. Löf, P. Madsen, P. Muiste, and L.-G. Stener. "Increased Forest Biomass Production in the Nordic and Baltic Countries — A Review on Current and Future Opportunities." *Silva Fennica* 50, no. 5 (2016).

Schwan, T. *Planting Depth and Its Influence on Survival and Growth.* Timmins, ON: Ministry of Natural Resources, 1994.

Sidders, R.G. *Bracke Seeding Rate Trial: 5th Year Results.* Thunder Bay, ON: Ministry of Natural Resources, 1993.

Sims, R.A., and K.A. Baldwin. *Landform Features in Northwestern Ontario.* Thunder Bay, ON: Ministry of Natural Resources, 1991.

Sims, R.A., W.D. Towill, K.A. Baldwin, P. Uhlig, and G.M. Wickware. *Field Guide to the Forest Ecosystem Classification for Northwestern Ontario.* Rev. ed. Thunder Bay, ON: Ministry of Natural Resources, 1997.

Smith, D.M. *The Practice of Silviculture.* 8th ed. New York: Wiley, 1986.

Smith, V.G., M. Watts, and D. F. James. "Mechanical Stability of Black Spruce in the Clay Belt Region of Northern Ontario." *Canadian Journal of Forest Research* 17, no. 9 (1987): 1080–91.

Stiell, W.M. "Growth of Clumped Vs. Equally Spaced Trees." *Forestry Chronicle* 58, no. 1 (1982): 23–25.

Symons, E. *Natural Regeneration of Hardwood and Softwood Tree Species Following Full Tree Harvesting in Northwestern Ontario*. Sault Ste. Marie, ON: Canadian Forestry Services, 1996.

Thorpe, H.C., and S.C. Thomas. "Partial Harvesting in the Canadian Boreal: Success Will Depend on Stand Dynamic Responses." *Forestry Chronicle* 83, no. 3 (2007): 319–25.

Tikkanen, Olli-Pekka, and Irina A. Chernyakova. "Past Human Population History Affects Current Forest Landscape Structure of Vodlozero National Park, Northwest Russia." *Silva Fennica* 48, no. 4 (2014).

Vaillant, John. *The Golden Spruce: A True Story of Myth, Madness and Greed*. New York: W.W. Norton, 2006.

Wagner, Robert G., and Stephen J. Colombo. *Regenerating the Canadian Forest, Principles and Practice for Ontario*. Markham, ON: Fitzhenry and Whiteside, Ontario Ministry of Natural Resources, 2001.

Wolf, Steven A., and Eeva Primmer. "Between Incentives and Action: A Pilot Study of Biodiversity Conservation Competencies for Multifunctional Forest Management in Finland." *Society and Natural Resources* 19, no. 9 (2006): 845–61.